11393

11468

omes

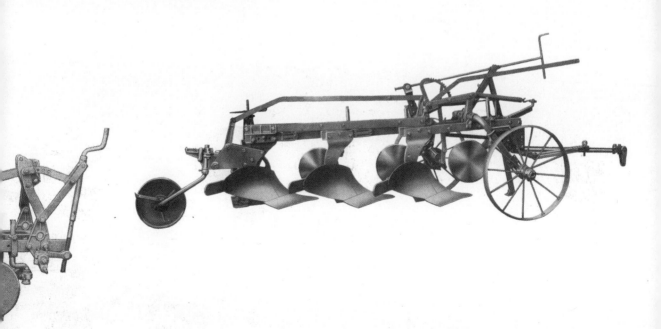

Ransomes
and their
Tractor Share Ploughs

Ransomes
and their
Tractor Share Ploughs

Anthony Clare

Old Pond
PUBLISHING

Published by

Old Pond Publishing
104 Valley Road, Ipswich IP1 4PA, United Kingdom

www.oldpond.com

Cover illustrations:
Front cover (top) *A No. 9 RSLD awaits delivery to a customer from the home warehouse.*
(bottom) *A TSR300D four-furrow reversible under test.*

Back cover *A TS59 at work near Ipswich - fine autumn weather,*
dry conditions and the soil turning nicely.
The rather formal headwear would not have been the norm for the
average ploughman, indicating that this is one of Ransomes' staff on trial work.

Edited by Julanne Arnold
Cover design and book layout by Liz Whatling
Printed and bound in Great Britain by
Butler & Tanner Ltd, Frome and London

Contents

Preface 6

Acknowledgements 8

Chapter One Ransomes Sims & Jefferies – the Company 9

Chapter Two Ransomes Sims & Jefferies – Plough Builders 12
Plough building at Orwell Works – Colour – Wartime production –
The new works at Nacton – Promoting plough sales.

Chapter Three Breaking the Ransomes Code 24
1900-1926 the alphabetic era; 1927 - 1930 the Trac era;
1931 - 1971 the "TS" era; Final designation –
The export ploughs – Components.

Chapter Four From Horse to Tractor 28
Popular early ploughs – Digging ploughs – Special purpose ploughs –
Twentieth century development – The early tractor ploughs.

Chapter Five The Heyday of the Trailing Plough 38
The introduction of self lift – A revolutionary design – The Motrac family –
Ploughs for special purposes – Ploughs for the market garden –
Ploughs for home and export – The final years.

Chapter Six The Ford Ransomes Era 55
Ford's plough manufacturing – The FR mounted plough –
The FR mounted reversibles – FR plough development in the 1950s –
Ploughs for the Fordson Super Major.

Chapter Seven New Theme – Finale 81
New Theme ploughs – postscript.

Glossary 88

Classification of Ransomes Tractor Share Ploughs 90

Index 93

Preface

Some years ago I purchased an old Ford Ransomes mounted plough to use with a vintage tractor in local ploughing matches. Like so many old ploughs, it had bits missing and new parts had to be obtained, but finally it was restored with YL bodies and has now been used for a number of years. The plough was a three-furrow Ford Ransomes PM and some years previously I had acquired some Ford brochures describing implements to go with the early Fordson Major including one for this particular plough. That brochure stated that the plough was "manufactured by Ransomes Sims and Jefferies Limited of Ipswich in association with Ford Motor Company Limited of Dagenham". On that basis I thought that I had a Ransomes-built plough, but I had some doubts. Having studied ploughs, and Ransomes' in particular, for a number of years I realised that there were some features on the PM which were unlike anything that one sees on a Ransomes plough and I resolved that at some time I would try to solve this "mystery".

I knew that there was a substantial amount of information about Ransomes at the Rural History Centre at the University of Reading, and in November 1997 I duly made a visit there to do some research. I was given a number of books, the most significant of which was a large black ring binder, which comprised an index of all the Ransomes archive material at the Centre, and some large leather ledgers, known as "Marks" books, which catalogued complete ploughs, mouldboards and sundry parts. It was noticeable, however, that much of the information only extended to the pre-Second World War period, although there was limited information relating to the post-war era.

I started to take notes of what I found because, although not directly related to the PM plough, there was much information relating to Ransomes and their products which was of considerable interest. The first visit flew by and I knew that I was going to have to make a further visit in order to provide an answer to my initial query. Next time I found myself looking at volumes of press cuttings, bound volumes of products manufactured by the company and large albums of photographs. As I delved further, some real "gems" appeared in the form of photographs of all sorts of interesting items, and the notes that I produced started running to a considerable number of pages.

I made further visits and I became more scientific in my research. I studied the bound volumes of manuals, sales literature and other details, and made notes of all of the tractor share ploughs that I could find. It was a bewildering task and I amassed lots of information, much unrelated and difficult to follow as to chronological development, but slowly a pattern began to emerge and I gradually compiled a list of the various types of ploughs, their bodies and other details. The research began to resemble a detective investigation, with odd pieces of information found, gaps required to be filled and various clues. As the months went by, the pieces of the jigsaw puzzle were slowly put into position.

The initial mystery regarding the PM plough remained, however, and because the Ransomes archive had not provided the answer I approached the Ford Motor Company at their Leamington plant, which I knew had been involved in some plough production. An article was put into their staff magazine which resulted in a number of interesting letters from staff who had worked on the ploughs, and this was then followed up by visits to various individuals in 1998. It also culminated in a visit to the plant, where I met their archivist who had written a book some years previously when the plant had been in the ownership of Ford for fifty years. She had a number of photographs of various activities at the plant, some of which showed the manufacture of ploughs, including the PM. Further research proved conclusively that rather than being manufactured by Ransomes as Ford's own literature had indicated, the PM was actually made by Ford, and Ransomes' input was relatively modest.

Although I had amassed a great deal of information on Ramsomes ploughs, I was left with a considerable number of queries about which I would need assistance from somebody who had worked for the company. I was put in touch with Fred Dyer, who was formerly the Plough Works Manager and is a mine of information about the company, and I went to see him in the autumn of 1998. I had several pages of questions to put to him, and more and more questions were answered and I was left with many fewer items to research. Research continued throughout 1999 and in the middle of that year I started to collate the two large ring binders, which were full of information, and write this book. It has been a daunting task. The huge amount of information available has had to be distilled into something which is not excessively long, but interesting to read. Some mysteries remain, but not many, and I think it unlikely, given the number of people that I have consulted, that some of these points will ever be answered.

I know that some of the archive information is incorrect, and I do not doubt that readers may be able to state that a particular body or part of a plough was fitted with something different from what has been stated as fitted in the works. Agricultural dealers doubtless undertook modifications to suit local demand and I have seen, and indeed owned, a plough so adapted.

One final point: only ploughs of Ransomes' own manufacture have been included, except those made under the agreement with Ford. Ransomes' acquisitions of companies such as Howards of Bedford and Hornsby of Lincoln are excluded, largely because they continued with their own production and sales, albeit with assistance for some manufacture from Ransomes.

ANTHONY CLARE
June 2001

Acknowledgements

Over the last two and a quarter years in which I have been undertaking this project, I have had assistance from a considerable number of people and I would particularly like to thank Dr Jonathan Brown at the Rural History Centre at Reading University, who has been my mentor and guide in researching the volumes of information entrusted to the care of the university by Ransomes, and Caroline Benson for organising the copying of the photographs.

The majority of the photographs in this book are reproduced by courtesy of the Rural History Centre, University of Reading. Others were supplied by the Ford Motor Company and Kongskilde or from the author's collection.

I would also particularly thank Fred Dyer, formerly Plough Works Manager at Ransomes. In many ways, this is his story and that of his colleagues too numerous to mention, who laboured on the shop floor and elsewhere in designing, developing, making, testing and selling the ploughs. I would like to thank his former colleagues, Geoff Teague, Paul Warnes and Chris Brown, all of whom were at Ransomes. I would also like to thank Tony Jennings and Bob Rendle, both of whom continued to work for the firm under its new ownership by Textron at the Nacton site at Ipswich.

I would also like to thank Charles Halliday of Kongskilde, whose early evening phone call one day in the spring of 1999 came as a ray of sunshine at a particularly depressing stage of the proceedings. A number of important documents that had not been deposited with the archive at Reading were found in the possession of that company in Norfolk, and he and Steve Smith, the Company Secretary, were most helpful in letting me have access to this information at a critical time.

Turning to the Ford Motor Company, I would like to express my thanks to Ralph Mills, the Personnel Manager at the Leamington Plant, who allowed me to visit and inspect the archive which is ably managed on site by Jackie Koord. Mr Mills was instrumental in letting me make a plea for help in the company magazine and that resulted in further assistance from Bill Lees, Jack Cox, and Ted Skinner MBE, all of whom provided invaluable information about plough manufacturing at Leamington.

I am much indebted to Stuart Gibbard for his help and encouragement and also, for guiding me to John Foxwell, who was executive in charge of the design of the Fordson Major and Dexta tractors from 1955 to 1962. Earlier in his career he had been involved with plough design and he has been most helpful in providing additional background information and he also put me on to Roy Keeble, formerly at Ford, who has provided similar assistance.

Grateful thanks are due to John and Anne Carter, particularly Anne, who has undertaken a considerable amount of typing and whose association with the project coincidentally led to some interesting information coming to light about Ransomes' involvement in competitive ploughing. I would also like to thank another former secretary Odette Barry, for helping to compile tables and other information essential for the basic research and Pam Pink for the plough drawing.

I am also grateful to Allan Condie for providing technical information from his archive which helped answer specific queries, particularly regarding the development of the early mounted ploughs. I would like to thank Tony Westbrook, a Sussex vintage ploughman of some repute, for his assistance and interest in the project, and Brian Bell, who has also laboured on a not dissimilar task, and provided useful information. A special "thank you" is due to my wife, Jeanette, who has had to put up with my hours of labour on the computer, and who I suspect thought there were other more pressing duties in the household on a considerable number of occasions!

Lastly, and by no means least, I should also like to thank, or should I say blame, Angus Montgomery of Billericay, another ploughman of no mean ability, who sold me that !!!! mounted plough in the first place!

CHAPTER 1

RANSOMES SIMS & JEFFERIES
– The Company

The business that was to become Ransomes Sims & Jefferies was started by Robert Ransome who was born in 1753. He was apprenticed at a Norwich ironmonger and subsequently set up a small foundry in the city. Along with other items, he developed and subsequently patented a tempered cast iron plough share.

In 1789 he moved to Ipswich, and started in business as a new iron and brass founder at a site known as St Margaret's Ditches. In 1803, while an iron plough share was being cast, a mould burst, causing the hot casting to fall onto an iron plate on the foundry floor. The underside rapidly cooled and was found to have become extra hard, a process that came to be known as chilled cast iron. It was further found that when in use shares cast in that way wore more quickly on the softer top than on the underside and were thus self-sharpening. This was an important discovery, and of potential financial benefit to Ransome, who patented the process in 1803.

He took out a further patent in 1808 for the standardised production of plough fittings which allowed iron ploughs to be repaired easily and put back in use by unit replacement rather than having parts repaired or made anew by a blacksmith. This allowed a much more rapid return to use of defective ploughs, saving time for the farmers. The process also permitted construction of a diverse range of ploughs which were developed from standard sets of components and was a feature of the company's manufacturing ability until the end of plough manufacture.

The vagaries of farming economics led Ransomes to diversify from early in the 19th century and in so doing, it again set the pattern for much of the company's activity until the late 20th century. One of their most significant diversifications was into the manufacture of grass cutting machinery, which began with the manufacture of an early Budding machine in the early 1830s. Manufacture of this sort offered a commercial cushion against the fluctuating fortunes of the agricultural trade.

By the mid 1830s the site at St Margaret's Ditches was becoming too small for the expanding business, and ten acres were purchased down by the River Orwell, which subsequently became known as the Orwell Works. In 1847 a new foundry was opened at the site and by the late 1930s, the works had grown to over forty acres. The new site, which had opened in 1841, was the location for plough manufacture until the mid 1950s. The Orwell Works was situated on the eastern side of the River Orwell and had a substantial quayside frontage on the south-eastern side of the town.

By 1843 the company had developed a new range of ploughs and one of these became known as the Yorkshire Light - YL - and won a prize that year at the Richmond (Yorkshire) Agricultural Show. It won further prizes, and as a plough body type it was developed continuously almost until the end of plough manufacture by the company.

In 1864 further fortune came the company's way when it developed a plough known as the Newcastle plough and won a prize at that location when the Royal Agricultural Society of England held a trial to test types of plough. Throughout the company's history they consistently produced products which were in the forefront of plough design and which were frequently used by winning competitors at ploughing matches.

Shortly after the turn of the 20th century, a new plough factory was opened at Orwell Works on what was known as the Knolls. The firm was soon to produce the first of its tractor drawn ploughs and continued to prosper until the First World War, when export trade came to an abrupt halt. This was a considerable blow to Ransomes, which had built up a substantial foreign trade in many parts of the world.

An aerial view of the Orwell Works showing the spread-out layout of the different departments. Plough assembly was undertaken in buildings adjacent to the quayside seen just above the lock in the bottom of the picture, with parts being manufactured in the forge and foundry shown to the left-hand side of the gas-holders. Further buildings for assembly and storage are located above and to the right of the quayside works.

Diversification into munitions and, in particular, aircraft manufacture, took place but after the war trading was difficult and a number of important export markets, particularly that with Russia, were lost. Further significant technical developments took place after the First World War, among them the introduction of the self-lift device, which will be described in detail subsequently. Trade continued to build up at a relatively slow pace until the Second World War.

Further disruption of trade took place in 1939, which was the year of the company's 150th anniversary. Another switch to munitions production occurred during the war, but a limited range of ploughs was manufactured in some quantity because of the importance attached by the Government to the production of food at home. Many areas of rural and, indeed, not so rural locations were put under the plough for the first time for this very purpose.

After the war, the problems of recovery were not so serious as had been the case after the First World War. World-wide trade improved more quickly and consistently and a further major development took place in the form of the agreement with the Ford Motor Company to jointly manufacture and market a range of ploughs to be used in conjunction with the new Fordson tractor which had been introduced in 1945. This arrangement was to continue for several decades.

Just after the war Ransomes purchased a site of 265 acres on the south-eastern outskirts of Ipswich at Nacton. In 1948 Ransomes opened a new foundry at the site and further development took place with the opening of a new plough works there in 1955. There was a further expansion of manufacturing and by 1968 the company had entirely transferred to Nacton from the old Orwell site.

The 1970s saw a period of considerable industrial upheaval caused by the oil crisis of that decade, the miners' strike and other industrial disputes which contributed to rapid inflation. All of those factors had an adverse influence upon Ransomes' fortunes, as indeed it had upon many other industrial manufacturers. It was also a decade when scientists began to change their ideas about the future of tillage techniques, and developments were afoot which led

A 1950s view of Nacton Works, the foundry being the group of buildings on the left, the first to be built on the site just after the Second World War. The main assembly building in the centre was constructed in 1955 and the plough department was the first to be entirely transferred from Orwell. The long bays to the side were added later for the manufacture of implements for the Fordson Dexta. The surrounding area was developed for other purposes by Ransomes and surplus land developed and sold or let to other users.

to the possibility of controlling weed growth with chemicals without the need for conventional ploughing. Larger implements had been developed to suit the higher horsepower tractors, and this led to a further decline in sales as changing techniques in cultivation caused a diminution for the sales of conventional mouldboard ploughs, even in the home market.

Ransomes at this stage were still manufacturing in a range of diverse fields. Apart from cultivation equipment, harvesting machinery in the form of self-propelled combine harvesters and sugar beet harvesters were being manufactured, together with electric trucks, material handling equipment, lawn mowers and turf care machinery.

The 1980s saw further regression as intense competition in the field of combine manufacture and electric and fork-lift truck production led to a decline in sales. With the prospect of substantial development funding being required in the near future and in a period of declining manufacturing in the United Kingdom, Ransomes concluded that it was best divested of these activities.

Later in that same decade, the same conclusion was reached with regard to the agricultural manufacturing business, including plough making, and in 1987 the business was sold to Electrolux of Sweden who set up a company called Agrolux to operate the former Ransomes business.

The new company initially traded with the Ransomes name alongside Agrolux, continuing to offer the current ranges of ploughs, and Ransomes continued manufacturing replacement parts, including the casting of shares for a limited period. The last plough share was manufactured at Nacton foundry on 31 January 1990, 201 years after Ransomes' founding, based upon that very activity.

The company was subsequently bought by an American industrial group called the Textron Corporation and the business which was acquired by them on 27th January 1998 now trades as Textron Golf, Turf and Specialty Products. Its premises are located in the 1950s assembly shop; the separate forge, foundry, spare parts and other buildings have been disposed of to outside interests.

CHAPTER 2

RANSOMES SIMS & JEFFERIES
– Plough Builders

Plough Building at Orwell Works

Ransomes' plough works at Orwell were located in three separate geographic locations. The assembly and machine shop was on a site known as the Knolls which originally was a swamp. The smithy was located next to the gas works and produced an extensive range of items for the plough works, as did the foundry which was located on an adjacent site; these shops also produced items for the company's other works such as thrashers, lawnmowers, etc. The drawing office was originally communal with other engineering departments, until a dedicated drawing office for the plough department was set up in 1919, located at the Knolls.

The opening of the Knolls coincided almost exactly with the introduction of the earliest plough dedicated for tractor use. Most of this earlier production, as we shall see later, was generally small scale, and the ploughs evolved from those which had been developed by the company for animal draught. While some models were produced in reasonable quantities, it was not until 1919 and the introduction of the RSLD/M and other similar ploughs that larger scale manufacture took place.

Ploughs at that time were generally constructed with plain section bars, with braces and stiffeners of similar form. These formed the basis upon which malleable iron skifes were bolted, the bottom section

A general view of plough assembly in the New Work Department taken in 1938 showing a variety of plough types under assembly with material on the right-hand side for manufacturing purposes. Not all of the items being manufactured are ploughs, as the plough department dealt with the manufacture of other tillage equipment such as cultivators, harrows, etc.

The PE516 was an experimental two-furrow digging plough, manufactured for the Forestry Commission, fitted with deep digging breasts with substantial clearance under the frame. Note the unusually large rear wheel. An early use of swan-neck beams, but still using lever adjustment.

of which was shaped in a shoe form and onto which the breasts were directly fitted. Ploughs were universally fitted with iron landsides and knife or disc and, later, skim coulters, with levers, wheels and sundries, all evolved from standard plough components which had been used for many years. The self-lift device for the RSLD/M was designed by the drawing office.

With the introduction of the Motrac a few years later, a radical change in the design of the plough frame took place, with the introduction of forged swan-neck beams of plain section, again fitted with stiffeners, braces, etc and other similar details of earlier ploughs. One innovation of the late 1930s that was seen on this plough, and became universal subsequently, was the introduction of the hub lift. Instead of an external quadrant being lifted on and off a sprocket directly fitted to the wheel, in the revised design the sprocket was enclosed within a case, lubricated with oil and used the engagement of an internal lever assisted by springs to lift the wheel in and out of work. This had the advantage that it was much less prone to becoming fouled by soil and was consequently subject to less wear.

By this time ploughs had adopted the use of a detachable frog upon which the mouldboard was fixed, but similar arrangements of knife or disc coulters were provided, with skims also being available if desired. Damage to the share and occasionally the frog had started to become a problem with the advent of more powerful tractors and of ploughs digging at deeper depths than had hitherto been impossible with animal draught ploughs. A sprung hitch was therefore developed that was fitted to the drawbar to prevent shock damage, and also to allow emergency release of the plough in the event it became trapped on an underground obstruction.

Some ploughs were built with frames incorporating what was known as the Demon or double lemon steel profile. This was an "I" section frame made with steel of high tensile form, with a vertical bar formed between lemon-shaped flange sections at the top and bottom. For more strenuous conditions, and where deep digging ploughs were used, these types of steel bars were found to be particularly suitable, being relatively light in weight. They also included similar arrangements of braces and ties, and lateral stiffeners were introduced across the top of the frame to give additional rigidity between the beams. Where plain section swan-neck beams were in use, several beams could be bolted together to give the required strength for the manufacture of heavy duty ploughs. Some of

The Demon section steel frame of a pre-war Multitrac. This was substantially stronger than other types of steel then in use for plough frames.

these early multi-furrow digging ploughs could only be used by crawlers, and consequently these were generally fitted with fixed drawbars as opposed to the adjustable hitches provided for the majority of wheeled tractors at the time.

Longer ploughs were invariably fitted with rear lifting wheels, but on smaller two- and three-furrow ploughs these were an option. For the short one- and two-furrow ploughs, rear wheels were not normally provided. All ploughs could be fitted with either disc or knife coulters, and those for the digging ploughs usually had knife coulters fitted to the frames. A pattern of disc coulter which was fitted with a cranked bar and a small skim coulter was introduced in the early 1930s, based on a design by one of Ransomes' agents, George Johnson, from Scotland. This came to be used almost universally for all general-purpose and semi-digger ploughs until the 1960s.

Plough design could vary significantly between similar types, and a typical example was the pre-war No. 4 Midtrac, which was a heavier duty general-purpose plough than the Motrac, and was provided with both a lever and a screw for adjustment purposes and plain section swan-neck beams with no rear lifting wheel. The post-war TS45 variant, by contrast, had screw controls, a rear wheel lift and plain frames, which were fitted with detachable legs.

The mounted plough era saw a continuation of the same design features, with either plain rectangular section steel frames, fitted with bolted malleable iron legs, or forged steel swan-neck

The standard disc coulter and skim attachment introduced by Ransomes was developed in the early 1930s from a design that had been adapted by one of their agents, George Johnson. Ploughs and attachments of all sorts were individually adapted for use in local conditions by agents and farmers, and Ransomes absorbed some of these changes and put them into production for their own purposes later.

RANSOMES IPSWICH
SINGLE ARM DISC COULTER
(CASE HARDENED BUSH)
WITH ADJUSTABLE SKIM

An early TS59 fitted with Epic bodies seen at work in the Ipswich district. Note the spoked depth wheel carried on from trailing plough practice.

beams. The double lemon section steel was not used, however, for any of the mounted ploughs. Early mounted ploughs were fitted with a standard spoked depth wheel of a type that Ransomes had used for many years, but after taking over the production of mounted ploughs from Ford, Ransomes bought the same pressed steel wheel manufactured by Sankey of Wellington in Shropshire as Ford had done before.

Control handles were made with steel cranked arms of round shape, with varying positions being offered to facilitate adjustment while working, depending upon the type of tractor to which it was fitted. An early problem was that because of the differing seat positions on various tractors, handles needed to be in a convenient position and flexibility had to be introduced in the headstock arrangements to provide optional positions.

Further changes took place with the introduction of the reversible TS50/51 mounted ploughs, and special high tensile steel, giving strength relative to weight, and which had been developed by Ford, was used for the frame. This was particularly critical, bearing in mind the substantially greater weight compared to a

conventional plough which was carried on the lift arms of the tractor. These changes also involved the use of both right- and left-hand bodies and mouldboards, frogs etc., which all had to be of reversible form, with disc coulters, skims etc., also of right- and left-hand pattern. Pneumatic tyres were offered as an optional extra, as they had been for one or two trailing ploughs.

The introduction of pre-set hydraulic linkage control on the Fordson Diesel Major in 1955 saw the start of the elimination of the depth wheel altogether. Further significant change occurred a decade later when rolled hollow-section steel started to be used for the frames, and this feature continued as a standard pattern for all subsequent plough designs and led to the complete elimination of previous types of frame arrangements.

As the reversible plough became dominant in the 1970s, plough bodies became larger, and this decade saw the elimination of many of the earlier types. Differing forms of skims were developed to take account of the higher speeds that came to be used for ploughing, and by now the majority of ploughs were digging at widths and in numbers of furrows unthinkable half a generation before.

Hydraulics started to play a further role in the operation of the plough with the introduction of the rear wheel lift on the TS78 semi-mounted plough. Later reversible ploughs started to use hydraulics for the turnover mechanism for the larger ploughs, and this became a standard feature with the later ranges. Trip legs, sprung or hydraulically actuated, were introduced in order to prevent damage to the bodies and by now the distance between legs was much greater than had been seen hitherto. Shear bolts had been used for similar protection purposes previously.

Mouldboard making in the early years was centred on the use of wrought iron that had been subject to a chilling process in the manufacture. This imparted a hardness to the surface, which was also durable and developed a polish which was beneficial in preventing sticky soils from adhering to the face of the mouldboard. The difficulty of producing iron or steel plate that was capable of turning a variety of soils without their sticking to the mouldboard had been a problem for a considerable period of time.

In 1830s North America, two Illinois blacksmiths, John Lane and John Deere, had both independently experimented with ploughs made of saw blade steel. John Deere subsequently became a household name in the field of agricultural engineering, but not so John Lane, who remains largely forgotten, although he is thought to have first developed a three-ply steel. These ploughs were found to be strong, flexible and durable, comprising fine grain steel which was ideal for use as mouldboards. This material, which came to be called Kristeel was commercially developed in the United Kingdom and in the early 1920s, Ransomes produced mouldboards from it. This was a three-ply steel of sandwich construction faced with a wearing surface of highly durable steel that was capable of becoming finely polished and ideal for use where sticky land conditions were encountered. The centre was a "softer" steel, with a harder rear plate making up the final ply. This steel became extensively used for the production of mouldboards in ensuing years and was produced by two companies, Rotherham Forge and Sheffield Forge.

In the early 1950s, a South Wales Company called the Steel Case Manufacturing Company of Tredegar perfected a means of heat treating a two-ply steel used for the production of discs for harrow purposes by Ransomes. The business owner, a Mr Thrupp, had been keen to produce mouldboards for Ransomes, but after some experimental production the company found that the boards were not of sufficiently good quality for commercial use. The business relationship between the two companies continued however, and in the early 1970s Ransomes took over the company

A No. 2 RSLM, the first to be fitted with Kristeel bodies in December 1922. This steel was specially developed to prevent heavy soils sticking to the mouldboard and was an advance on the wrought iron and steels that had hitherto been used. Note the arms for fixing the skims to the disc holder.

A 1938 view shows tractor plough manufacture on some scale, with ploughs being assembled by individual fitters, with what looks like a foreman supervising about ten men. On the right-hand side a small number of completed ploughs can be seen, although not yet fitted with coulters.

and rebuilt the factory. From the mid 1970s onwards mouldboards started to be produced by Steel Case, in carburised boron steel, and this material came to be used alongside Kristeel for mouldboard production until the end of plough manufacture in the late 1980s. Subsequently some of the manufacturing equipment, including the dies for forming mouldboards, was acquired by Dowdeswell from the South Wales plant, and they took over the production of the YCN bodies for their own purposes

Colour

We now turn to the vexed question of colour. Vexed because there seems to be a widely diverse opinion among those who restore Ransome ploughs as to what colour is correct. Part of the problem has been compounded by the modern description of colours such as Mounted Implement Blue and Trailing Implement Blue. In fact, the company made no such discrimination as far as colours were concerned and until the Second World War there was a universal scheme which applied to all Orwell-manufactured ploughs. The body, including the controls, was painted in what equates closely to what is now called Orwell or Trailing Implement Blue and the wheels were painted red, as were the backs of the mouldboards. Coach lining was applied on the frames and afterwards varnish was applied to the surface of the polished steel mouldboard.

The paint was initially provided by Docker Brothers, but during the Second World War, paint production in as wide a range of colours as before became more difficult. Some of the pigments used to produce the original blue colour were not available, and colours changed to a darker tone and hence evolved into what is now regarded as Mounted Implement Blue by some or Nacton Blue by others.

After the Second World War it became more difficult to find operatives to paint the ploughs by hand, and the time expended in lining and painting in differing colours had been dispensed with as an economy measure. Henceforth ploughs were universally painted in a single colour and were painted directly on to clean bare metal. The works at Orwell had built what was known as a flow-coating bay, in which a series of suspended angled nozzles connected to pipes applied paint to all parts of the plough as it passed through. Surplus paint was allowed to drain into a large containment tank and was re-cycled for use. The ploughs were then taken to a drying oven where they were baked, after which decals were applied to the frames with the "Ransomes" name and "England", including a stencil marking on the face of the steel mouldboards stating that they were of Kristeel manufacture. Small embossed metal plates were fitted to the front of the frame, identifying the plough mark and any additional descriptive letter.

Ploughs for some export markets were painted orange, again of all-over application.

The colour used by Ransomes for the post-war ploughs applies to later manufactured trailing implements as it does, of course, to the mounted type. Consequently, the post-Second World War RSLDs and Ms would have been painted in the darker colour and not the lighter hue that was used before the War. Hence the No. 13 and 15 models, the last produced, would have been so painted whereas those such as the No. 9, built pre-war, would be the lighter colour. The differing colours and style would apply to all the pre-war plough types which continued in production after the war. By a strange coincidence, when Ford began manufacturing ploughs under the terms of the joint agreement, the colour was relatively close to that used by Ransomes, although probably a little lighter, corresponding to the Empire Blue which was subsequently used for the new Fordson Major in 1952.

One frequently sees ploughs painted in the present era with orange depth wheels but there does not seem to be any evidence for this practice and it was never adopted by Ransomes. Both Ford and Ransomes fully assembled their ploughs, which were subsequently painted all over in one colour. The early Elite plough produced by Ford had been a darker blue, in fact similar to that applied to the E27N Major, and both the land and furrow wheels were painted in the same orange that was applied to the tractor wheels. It is possible that some early mounted PM ploughs had orange wheels which were painted by the wheel suppliers to give surface protection prior to delivery to the Leamington works, but according to information from some of the men involved in production at the time, that was not the normal practice, because ploughs were painted all over in one colour after assembly.

Wartime production

Manufacturing at Orwell had been diverse, in that a very large range of different ploughs and indeed other cultivating implements were manufactured. This led to a huge variety in terms of manufacture of all forms of components, and the assembly of these items must have required considerable organisation bearing in mind the constraints of the older turn-of-the-century building, and just after the Second World War Ransomes ended horse-plough manufacture. After the cessation of hostilities on the 22 October 1945, Ransomes produced the 30,000th Motrac plough. This was quite an achievement, particularly as during the

A war-time view of production showing Motracs on the assembly line. At least four women can be seen assembling both components and complete ploughs, an indication of the increased use of female labour to free men for other war work.

The use of Ransomes' manufacturing facilities for wartime production led to what were known as village groups set up to provide assembly in locations far from the factory. These could be in village halls, blacksmiths' shops and other similar locations and involved a wide range of unskilled labour. Here is a range of the type of products which were assembled by such groups and we can see a plough drawbar and rear wheel assembly.

war there had been constant problems finding sufficient manpower to allow maximum productivity to continue. In fact, large numbers of women were employed in the plough works, as indeed elsewhere in Ransomes and other manufacturing companies, in order to make up the shortfall of male employees. The Government had set a production target of 160 ploughs per week, but in practical terms this was only once achieved. A maximum of 600 employees was available "on the tools", which was a 100 per cent increase on the numbers in the works before the war. These were overseen by a works director who was in charge of overall production for the company and to whom four separate manufacturing managers reported, including one for the plough works. Beneath these managers production controllers, shop superintendents and foremen provided other managerial and supervisory line staff in charge of the operatives.

In order to provide additional capacity and to provide a degree of diversification in the event of damage to the factory, Ransomes arranged what were known as village groups to undertake limited production. These were centred in outlying districts, and workers undertook simple manufacture of small and even larger assembled components at home or in blacksmith shops or other appropriate locations. The company expanded these facilities and eventually items such as drawbars, rear wheel assemblies, handles, bearing assemblies and the like were all being produced. The numbers manufactured were probably relatively small, but no doubt comprised an important contribution to Ransomes' requirements.

Manufacture of an implement that was to see "front line" service occurred when Ransomes developed what was known as the Farmers Deck body. This was a large digger-type mouldboard that was mounted on a long jib, and two of these were fitted on the front of a Churchill tank. The intention was to literally "plough" mines to the surface of the ground and the tank was followed by an Equitine Cultivator whose tines assisted in exposing the mines for subsequent detonation. This allowed a safe passage for vehicles and personnel to follow in the rear of the track.

The war saw other changes to Ransomes' manufacturing activities. The company had retained vast numbers of patterns for ploughs from many years previously. In fact until 1944, when they were destroyed, patterns were available for the original 1843 YL plough as well as for large numbers of early and obsolete items. The company had been dominated by sales requirements for many years,

which had resulted in the availability of an enormous range of different ploughs, which must have caused the manufacturing department many problems. These had generally been introduced at the behest of the sales department and the Home Sales Director, Henry Deck, in particular. Some types produced sold in very small numbers, for example the TS24 Twinwaytrac, a two-furrow balance plough, was made for a single farmer who bought just five! Gradually production requirements took precedence over sales, resulting in a considerable reduction in available plough types. Throughout this post-war period, Fred Dyer, the plough department manager, and his staff spent some time each day destroying obsolete patterns and rationalising the range of available items.

The new works at Nacton

Ransomes' plough manufacture in the post-war era of was dominated by a major expansion of their manufacturing facilities. The new site at Nacton was developed initially with a foundry and later with an assembly shop where on behalf of Davey Paxman of Colchester the company assembled a batch of diesel engines, which were being supplied to the Royal Navy for a class of minesweepers. After the completion of this contract the building remained empty for a while. Plans were drawn up for construction of a new plough manufacturing building after Fred Dyer produced a scheme to show what was required for current manufacturing purposes. The directors were shown the proposals and gave approval for building to begin, and in March 1954 a contract was let to a local builder,

Cocksedge, for the construction of the new building, which was completed by December.

The new workshop comprised a large, modern steel-framed building, with up-to-date facilities for the production of ploughs and other cultivating implements. Three powered assembly lines were set up within the factory, and the former diesel engine assembly shop was converted into a machine shop. From 1955 onwards production of ploughs and other tillage equipment was gradually transferred from Orwell, and eventually all agricultural equipment manufacturing and other manufacturing activities of the company were transferred. In 1956 further substantial expansion took place, with the addition of extra bays on the main assembly building, which were erected for the production of a range of implements to be manufactured for the new Fordson Dexta.

Manufacture henceforth took place on a much more sophisticated basis. Instead of moving items along a fixed track by hand, electrically powered tracks were now available for the assembly of implements on a much larger scale. Batch production took place on the three main lines, and individual types were produced in batches with changes taking place on each line two to three times a week. This allowed up to 600 of any particular design to be produced on a weekly basis, although at one time the company had planned to produce up to 1000 implements each week.

Ploughs for export were sometimes built as what were known as knock-downs and packed as kits, to be assembled in a distant country. Ransomes had factories in Mexico and Colombia in South America, and some manufacture took place in Spain, the plant being subsequently taken over by Ebro. At the end of the Nacton production line, ploughs were removed and cleaned prior to passing through a flow-coat spray system and then stove baked as they had been before.

Plough design continued to evolve and each type was by now fully designed, whereas in the earlier years of manufacture, designs tended to evolve one from another in a way which did not necessarily require a high degree of initial drawing office involvement. Ransomes had the use of some fields near the factory where testing and development could take place, and throughout the history of the firm considerable experimental design work was undertaken. Geoff Teague went to Ransomes in 1944 and spent much of his time testing ploughs, including the Jumbotrac, Multitrac, the TS50/51 reversible, and the later TS68 reversible among others.

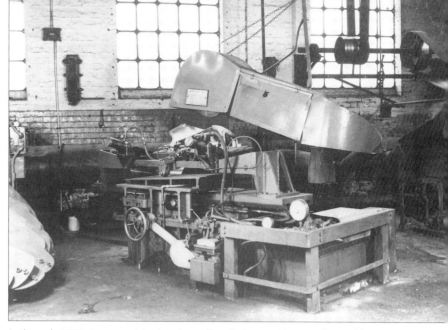

In the early 1950s Ransomes introduced a profile-grinding machine manufactured by Robertsons of Bedford for polishing mouldboards. Hitherto this work had been undertaken manually by boards being fitted to a barrow frame and passed backwards and forwards under a large grinding wheel. The work was laborious and was one of the tasks for which it was increasingly difficult to find workmen after the Second World War, hence the need to develop a machine for the purpose. Note the use of other belt-driven machinery on the right.

Ipswich and the surrounding countryside were unusual in having a considerable diversity in the types of soils in the fields; consequently ploughs could be tested in a wide variety of soil conditions, with the exception of rocky subsoils. Farms at nearby locations such as Playford, Akenham, Westerfield and Bucklesham were used for such trials and many of the photographs of ploughs at work in Ransomes' archive were taken at these locations.

The Nacton plant was fully automated and must have represented state of the art in implement manufacture when new. Here we see two lines, on the right-hand side on the powered track TS59s being assembled with, on the left, on the manual track, TS54s. Note the use of pneumatic tooling and other labour-saving devices.

Ransomes introduced an automatic flow coat painting booth for painting implements at the end of the line. Here is a two-furrow plough being lifted from the adjustable table which was capable of taking ploughs from either of the two main production lines.

Promoting plough sales

Ransomes also continued another practice with which they had been involved for some considerable time, and that was to give help to competitors in ploughing matches. There was enormous prestige to be gained by a winner using a particular company's products and Ransomes supported a number of ploughmen who competed at local, national and international level. This support extended to the adaptation of ploughs and equipment, and even late in the 20th century special bodies were developed for use by competitors. Two variants from standard ranges of later ploughs were adapted for use for competition ploughing, the TS86 and TS97.

Assistance from the company extended on occasions to advice to competitors during participation in the actual match, normally forbidden by the rules of the match or district. A Cotswold farmer recalls being "assisted" by a company representative with a special code of signals to indicate necessary adjustments to make the work look right. For example, a stick was usually carried, and tapping this stick on the ground with either the left or right hand meant the relevant body required lowering. Other movements of the hands, arms or legs indicated other adjustments, such as to discs or skim coulters, etc! Winning at matches represented not just prestige, but potentially extra sales, making such

Ransomes supported ploughing matches and particular participants, making technical advice and equipment available to promote their own ploughs. Here we see an RSLD ploughing in a high cut class at a match in the 1950s. Note the awkward, not to say potentially dangerous, position of the ploughman!

dubious practices worthwhile commercially.

Ransomes also promoted the practice of ploughing at a more general level, and from the middle of the century almost until the end of plough production, they produced a book called *Good Ploughing*, in which the liberal use of photographs and drawings taught farmers and novice ploughmen the correct settings for the plough and the techniques required for good practice.

The sales agents at home and abroad had a range of literature for the various commercial ploughs. In the early days this consisted of a single sheet with a simple lithograph illustration and basic information about the uses of the plough and the types of soil conditions for which it was suitable. Operators' instructions and parts catalogues were not produced until the early mass-produced ploughs were introduced and some special-order ploughs had no descriptive material at all. By the 1930s general plough brochures started to be produced to describe the rapidly expanding range of implements. Most plough types also had a parts book describing the components, which were usually illustrated. Later, the technical support brochures were divided between simple parts books and those that also contained information for the

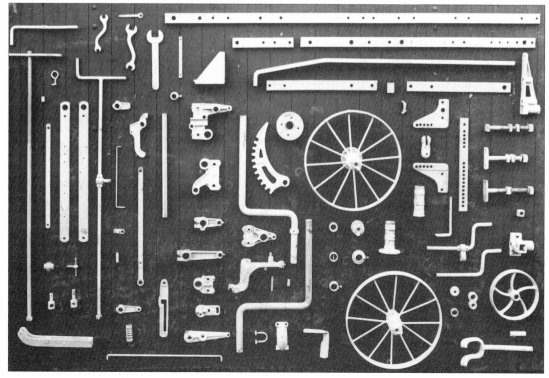

A further view taken at a Ransomes' ploughing match where a competitor with a two-furrow mounted plough is completing work around his crown. Note the substantial weight laid upon the frame of the plough.

operator on assembly, use and maintenance.

Photographs were taken in the works from overhead stands, the parts being mounted on dark wooden boards with the appropriate part number shown alongside. Complete ploughs were photographed in a variety of locations, in and outside the works, for both brochure illustrations and recording purposes. When the move was made to Nacton, the company built a studio for photographing its products, and items could be positioned with the use of a turntable with an overhead gallery, built for plan views. Within a few years, however, some of the implements became too large to manoeuvre in the studio and a return was made to outside photography.

Publicity brochures began to be printed in colour, usually with a second colour in addition to the black, and this was first introduced just after the Second World War. Some of these brochures were attractively designed in Art Deco style, but fully coloured brochures did not arrive until the 1970s.

The pre-war period saw the high point in brochure production, with Ransomes producing county and even dealer brochures detailing the implements that were made for particular areas. Other brochures show the vast range of spare parts and optional items of equipment available such as different mouldboards, coulters and wheel assemblies where, for example, the rear wheel lift was an "optional extra". Separate publications were available for the export ploughs and tillage equipment generally, and these were printed in the language of the relevant country. The later rationalisation of plough types greatly reduced the numbers of different booklets and most of these publications were printed by Ransomes in a special department.

Ransomes set out parts on shadow boards, with components painted in light colours, for the purpose of taking photographs for their illustrated manuals. Here is one of the illustrations for the RSLD/M Handbook.

CHAPTER 3

BREAKING the RANSOMES CODE

To the uninitiated, the descriptive letters given by Ransomes to their ploughs can seem baffling almost to the point of total incomprehensibility. The descriptions employed when tractor ploughs became separately developed from animal-draught ploughs were a continuation of that used for many years previously. Ploughs for the export market were generally given names, whereas those for the home market were sometimes initially named, but more usually took a combination of letters, and were subsequently named in the late 1920s. That system changed radically in the early 1930s with the introduction of the TS numbering system.

1900 -1926 - The Alphabetic Era

The earliest ploughs developed by Ransomes took alphabetical letters; consequently the first was called the A plough and so on through various letters of the alphabet. Later plough types were designated with names - and hence letters - relating to the areas with which they were associated or made, particularly when the types concerned became prizewinners such as the Y L Yorkshire Light and the R N Ransomes Newcastle. Incidentally there was a Yorkshire Heavy, the YOH, but it was not used as a tractor plough.

The early years of tractor plough manufacture were characterised by the use of names of various tractors with which they were used. For example, the letter M occurs frequently in early designations indicating that it is a motor (tractor) plough; and where a plough was made for a specific manufacturer such as Saunderson, the letter S became incorporated in the designation. Therefore an RSML is a Ransomes Saunderson Motor Light.

For home production, the earliest bodies in general tractor use were the YL and the TCP. The YLTM of 1909 stands for Yorkshire Light Tractor Multiple and RMTM stands for Ransomes Motor Tractor Multiple, with RMTD for Ransomes Motor Tractor Double.

The letter M for Motor disappears within the first ten or so years of production, as do the separate designations of body types such as YL and TCP, which were frequently added to a plough designation. The RYLT stands for Ransomes Yorkshire Light Tractor, and TCP on a digging plough would refer to the Titt Chilled Plough body used instead of the YL.

Letters contained within the plough descriptions could also refer to the number of furrows the plough could turn. D stood for double, and M for multiple. Multiple ploughs were three furrows and over; consequently the RSLD stands for Ransomes Self Lift Double; RSLM for Ransomes Self Lift Multiple and the M classification was used mainly for ploughs made until the mid 1920s.

In the early years the company decided that the letter X should designate an export plough. This was not a universal application, as will be seen later, although X for overseas trade appeared on an intermittent basis until the Second World War.

Versions of the early ploughs until the mounted plough era took numbers as a means of sub-identification. Consequently the RSLD/M plough took up to fifteen numbers, although Nos. 11 and 14 were not put into production. Re-lettering of ploughs could occur seemingly at the whim of the Sales Department, which would re-designate a particular plough type with a different number even though there was no evidence that the plough concerned differed in any significant form from its predecessor. For example, the RSLD/M No.5 was changed to No.7, and the only difference was that the 7 could be extended to take a fourth body if required.

The semi-digger and early digger types in the 19th century used breasts that incorporated chilling in the manufacturing process, and therefore those types of bodies normally incorporated the designating letters DC Double Chilled, in the title. The curious aspect of that description is that it is not possible to double chill a plough breast except by means of special techniques which are complex to implement and not undertaken in mass production.

In the majority of instances where ploughs were developed by Ransomes for both home and export trade, the letter R for Ransomes appeared in the title. Where plough types were developed by others, but subsequently adopted by Ransomes, the letter R was not always used. Consequently the TCP - Titt Chilled Plough - was a development by a Mr Titt of Warminster in Wiltshire in the late 19th century of an earlier Ransomes chilled plough which had not proved altogether successful in use. Subsequently the Titt body came to be used extensively until the Second World War. The SCP Steel Chilled Plough and LCP Lincolnshire Chilled Plough were other dominant types used in the early years of the trailing plough era, but they gradually came to be supplanted by the Irish variant, which unusually took the letter I for Irish before R for Ransomes in the designation. Hence IRDCP stand for Irish Ransomes Double Chilled Plough. Where the letter T is shown after the designation, it stands for tractor, indicating that the body was developed for that particular purpose. When Ford developed a lighter version of the IRDCPT for mounted ploughs after the Second World War, this became known as the FRDCP Fordson Ransomes Double Chilled Plough, even though chilling was no longer used in the manufacturing process.

1927 - 1930 - *The Trac era*

In 1926 change occurred with the application of a series of names with the final letters TRAC for a range of ploughs made until just after the Second World War. For example, the first mounted plough introduced in 1927 was called the Weetrac, presumably because the plough was small and light in weight, being provided with a mechanically operated lift and bolted on the back of the early Fordson F and International tractors. Of these similarly named ploughs the most notable is probably the Motrac - Motor Tractor, followed by the

Midtrac and Multitrac, the latter a three- or four-furrow version with the five-furrow Quintrac and six-furrow Hexatrac continuing the series. When a new range of digger bodies was introduced they were initially of one or two furrows only, that being all that could be hauled by the early tractors, and they consequently became Unitrac and Duotrac.

The designation for these digger ploughs used the same system as for the general-purpose ploughs, where the descriptive letters relate to specific plough types. Hence, for example, the post- war DM digging body was derived from the DeMon plough. What initially were regarded as deep digging ploughs, later called digger types, originated with the Unitrac plough from the early 1930s, and the body was classified UN. This body type was slightly varied with the introduction of the Duotrac plough and then took the letter D instead of N, becoming the UD. A further version based upon a plough body, the Guerri, manufactured by the Italian plough company Martinelli, then took G as the second letter after the U. This rule with regard to the descriptions, however, did not apply universally, which causes a further problem in working out the origins of the ploughs. For example, a derivative type of Unitrac body known as the Swampland became designated SL. The letter M was sometimes added as a suffix letter after the introduction of the mounted ploughs, to show that the body type differed from that fitted to the trailing type, hence the letters UDM rather than UD. This probably indicated a detail difference in construction in the two types, because the suffix letter M was not generally used on other types of bodies fitted to mounted ploughs.

A further descriptive word introduced in the pre war period is the word Major. A plough incorporating that name was generally one that was developed as a stronger version, with either additional under-beam clearance or with strengthened frames for particularly adverse soil conditions. The converse also applied with the description Minor, which was applied to lighter adaptations of ploughs that were already of robust construction, but where there was a need for a particular type to be produced for work of less strenuous sort, and these were built with lighter frames and details.

Development of individual ploughs continued to use numbers for additional classification, but what confuses the issue further is that some plough types did not necessarily take sequential numbers when further variations were introduced. When a variation of the No. 5 Multitrac was fitted with alloy axles, it did not become the No.6 but the 5A. Similarly the No.6 was fitted with strengthened wheels and became 6A and not 7. Further variations of the mark continued, with sub-designating letters A and B being subsequently introduced for the No. 9 model. Tracing these individual plough types, both as to number and additional identifying letters, is a complicated matter, and there is no simple means of identifying this multiplicity of changes from information contained within the archive at Reading.

1931 - 1971 - The "TS" Era

In 1931 the whole system changed again when Ransomes began to classify all their implements with prefixed letters. This was when the TS system of classification was introduced, TS standing for Tractor Share. Letters of a similar sort were used for other implements, for example, TD for Tractor Disc, C for Cultivator, H for Harrow, and P for Potato applied to implements for those cultivation purposes.

The first plough to be so designated was the TS1 Solotrac and the appendix at the back of this book gives the complete list of TS numbered ploughs. Not all plough types utilised the system, and some that had been introduced before 1931 continued to carry their old letters or names, although confusingly, some ploughs from the same era were subsequently renumbered according to the new system after the Second World War, because presumably they were going to continue in production. For example, the Motrac introduced in 1927 became known after 1945 as the TS43, with the TS44 being the Litrac, an export variant. After the Second World War however, the TS number became the dominant description. The plough name was usually shown on publicity and

operators' literature in the 1930s, but the TS number was not normally a prominent feature of the plough description in the way that occurred after the introduction of mounted ploughs, and sometimes was not shown in brochures or handbooks at all. The TS designation for home-produced ploughs ran to just over 100.

A larger incremental gap in numbers had been previously planned with the introduction of the Fordson Dexta, and a group of ploughs had been developed which were originally to be classified as the TS2500 and onwards, but for some reason this changed to TS1013 to 1016, comprising two one-way and two reversible ploughs. Why the numbering started at 13 and not at 1000 is a mystery and in fact Ransomes had initially allocated TS numbers 75, 76 and 77 for three of these ploughs, the general-purpose variant being based on the earlier TS54.

Additional classification and development of individual ploughs after the Second World War took letters rather than numbers, which had been used for such identification before that period. Some of the major plough types were extensively developed. The TS59, for example, had over thirty versions, although not all were put into production, and a few were additionally given numbers and letters for identification purposes. Other plough types such as the TS4 Jumbotrac also took a substantial number of differing letters to indicate the specific widths for which it was designed to plough and the types of bodies that were so used. When one takes into account the requirement to strengthen differing types for more strenuous work, a large number of variations and hence descriptive letters were involved. This plough was also interesting in that it was the only one of the entire series that had two separate names incorporated within the same TS designation. The TS4 was called the Giantrac and was renamed the Jumbotrac for the TS4D and then renamed Giantrac again for succeeding models. It is hard to understand the logic behind this, particularly when one considers that an export plough such as the Duratrac, of which three variants were introduced in 1949, was given a different TS number depending upon the number of furrows. Consequently the three-furrow version was the TS47, the TS48 the four furrow and the TS49 the five.

The jointly marketed FR Ford-manufactured mounted ploughs retained the designations given by Ford, which are described in Chapter 6, and were not given a number in the Ransomes system. Ford Ransomes implements were given Ransomes numbers and letters when the company took over their manufacture at Ipswich in 1955.

Final designation

In the 1970s the system changed again, when the letter R was used to designate reversible ploughs which were then starting to be sold in much greater numbers. Subsequently the numbering system jumped in hundreds, with the introduction of the TS200 Spaceframe non-reversible ploughs and the TSR100 and subsequent TSR300 reversible range. Note the return - an isolated case - of the use of a name in the description!

An earlier practice of Ransomes in using names of other body manufacturers occurred late in the 20th century with the introduction of the TS97 Ransomes Bonning which was a match version of the TS90 mounted plough taking Bonning competition bodies.

The Export ploughs

Names were in general use for the export market, at the request of local agents, and one of the first was a plough produced for Spain in 1909 called the Rey which is Spanish for king. Ransomes subsequently produced a number of export tractor ploughs with titular names such as Dictator, Autocrat, Consul, Vice Consul, Ambassador and Proconsul. These titles were not exclusively used, however, because when the letters TRAC were introduced for ploughs for the home market, Ransomes added prefix names such as Junotrac and Supertrac, in addition to the digger ploughs, such as Solotrac Unitrac etc., which had been developed for both home and export use. Even with the TRAC suffix naming system in place, Ransomes still continued to produce ploughs

with titular names, and for example, the Marquis of 1936 reverted to the earlier naming regime.

In the post-Second World War era, bird names began to be used for export mounted ploughs. The earliest of these was the TS54 mounted plough, which was initially introduced for the Ford 8N AN tractor in the export market and became known as the Robin. When this plough and subsequent variations were sold in the home market, the name Robin became generally applied, although it was never officially used as such by the company, being referred to as the TS54. Other export ploughs also had bird names which were not used at all in the home market, and it is a fair bet that many ploughmen would not have known that their TS59s, 64s etc. were so called. For example, the TS55 single-furrow digging plough had the export name Falcon. The name Raven applied to the TS59 and Vulture to the TS64. The TS68 reversible took the name Swift, the TS73 four-furrow semi-digger took the name Heron and the TS74 single-furrow reversible with a digging body took the name Swallow. Subsequently, Robin was used for the TS1013 manufactured for the Fordson Dexta, and the Tern was the export name applied to the TS1014.

Components

Ransomes used a large number of components in the manufacture of an individual plough. Letters and numbers were used for identification purposes, and some of the prefix letters were part of a system that had been used by the company for many years. Other identification marks related to the plough on which they were first used, and this designation would then apply when used on future models taking this same part. Reclassification subsequently occurred for some items, which can make identification difficult when trying to find a part for an old plough. In some cases the relevant parts book gives the new designation to later produced items. Similar marks identify individual components, but these are so numerous that their identification is outside the scope of this study into the naming of plough types.

CHAPTER 4

From HORSE to TRACTOR

When the tractor share plough was introduced at the turn of the 20th century, Ransomes had already been manufacturing ploughs for a hundred years and they probably did not regard this new development as anything more than a part of the evolution of the plough. Indeed, there may have been many at the time who thought that tractors and their ploughs were but a quaint experiment and that the use of the horse in agriculture, and indeed elsewhere such as in commercial transport, would remain supreme for many years.

Since the Middle Ages, English ploughing had developed into a craft, the practice of which required skill and intellect. Not all farm labourers were able or "permitted" to plough, because rules introduced in that era had become entrenched in agricultural practice and those rules determined who could and could not plough. Those notions seem odd to us at the beginning of the 21st century, when ploughing appears to many to be an activity in which technology has dictated that relatively little skill is required, or indeed applied, by large numbers of those who work as tractor drivers.

In the 19th century the practice of ploughing was aided by scientific development in materials and manufacturing practices, in which Ransomes were a leading company. Ploughing techniques became highly developed, aided and abetted by ploughing matches which had been in vogue since the 18th century. Skilled ploughmen were much sought after by manufacturers to show off their products, and indeed were well paid for their services. Thus, throughout the century what is known as the English style of ploughing developed, aided by particular designs of plough which will be discussed later.

The types used depended upon local soil conditions, regional preferences and what one might suppose would now be classed as irrational sentiments with regard to what was thought to be best. The substantial enclosures of the open fields had previously necessitated a technique that would not have applied, for example, in the veldt of South Africa or the savannah of the Argentine, to name but two export markets where Ransomes sold their products. In less developed and sparsely populated countries technique was of less value than the basic requirement merely to turn over as much soil as possible for cultivation purposes. Furthermore, soil types did not always lend themselves to laying down the traditional box section furrows, although in some areas, principally New Zealand and parts of Australia, the English style was considered to be of importance.

In the 19th century there was a proliferation of plough types for varying local soil conditions, and a quite remarkable number of different plough designs had evolved. Basic types were frequently adapted by local agents, large scale farmers and agricultural contractors, who would make changes to a plough breast that would be returned to the factory and subsequently manufactured by Ransomes as a specific plough type. Frequently the letter designating the party's name would become incorporated within the plough title.

Popular early ploughs

In the 19th century three particular plough types were made famous by Ransomes and were to continue into the next century. They became extensively developed and continued in production for the first fifty or so years of tractor plough manufacture. Indeed, one of them, the YL pattern, was not discontinued until almost the end of Ransomes' plough production in the 1980s.

The earliest of these plough types was what was known as the Yorkshire Light - the YL - which was first manufactured in 1843 and demonstrated at the Richmond (Yorkshire) Agricultural Show in 1844. This plough became very successful, winning a prize

The YL163 and other bodies for Kent were sold into a fairly small market and were typical of Ransomes' policy of manufacturing for such customers. This policy was abandoned shortly after the widespread introduction of the mounted plough.

at that show, and then another prize from the Royal Agricultural Society of England in the same year, along with other acclaim.

This plough type was developed from the original pattern YL1 in 1844 until 1953 when the final variant, the 187K, was produced for Stormont Engineering of Tunbridge Wells, Kent, for sale as a locally used implement.

Plough breasts were initially manufactured in chilled wrought iron, but steel came into use at the end of the 19th Century and in 1923 a steel called Kristeel was introduced for the first time. This was a fine-grained steel capable of turning sticky soils. The use of the suffix letter K after a mouldboard indicated it was made of this steel. Among the most used variants of the

marque for the tractor plough were Nos. YL79, 136 and 163, the latter of which was a special body produced for Kent.

The YL165, which was to become extensively used, was first introduced in February 1920, with further versions being produced for horse work and for special match purposes throughout the 1920s. In 1933 a variant on the YL165 was adapted by the firm Booth MacDonald, an agent of Ransomes in New Zealand, and became classified as the YL183. This was of a deeper form than the 165 and was subsequently manufactured on a large scale. Something like twenty

A considerable number of shares were available for use with the YL body. Here is a selection from the mid 20th century.

different shares were available for the YL body, more than for any other type used for tractor work. This plough type sold almost exclusively in the home market and in Australasia, where the English style of ploughing was dominant.

In 1864, further success came to Ransomes with the design of a plough which subsequently became known as the Newcastle type, winning a prize at the Newcastle trials in that year. Four principal variants were introduced for different types of soil condition, with the RND, first manufactured for light soils, being introduced in March 1872. The RNE variant was for light and mixed soils, the RNF for mixed and heavy soils and the RNG for deep ploughing in heavy soils.

Something like 213 types of this plough were introduced, a number of which were not put into general production, the last being introduced in November 1925. By the time that the tractor plough had become developed, this body type was losing popularity for general work and only a small number were introduced for use with tractor ploughs, other types, principally the YL, being more popular for what was known as general-purpose work. As a marque however it continued, being particularly popular for use in competition ploughing. About seven different points were made for use with the breasts used for tractor ploughing. The Newcastle sold in similar markets to the YL, where soil conditions suited its use.

Thirty years after the Newcastle trials, further acclaim came to Ransomes with the introduction of the IRDCP - Irish Ransomes Double Chilled Plough - at the Dublin Horse Show in 1894. This body type was probably the most significant in tractor plough development and became known as the semi-digger type. It was suitable for arable cultivation in a wide range of soil conditions, producing a furrow that more rapidly weathered to a flat seed bed than the general-purpose types, and thus requiring fewer subsequent tillage operations. This type of plough was particularly suited for arable ploughing, turning it into the dominant plough type during the middle of the 20th century. There were only a few versions of

this plough for tractor use, the most significant being the IDCP41, which was introduced in December 1930, the IDCP38T in March 1935 and the IDCP39 in September 1936.

The frog was found to have a weakness, particularly at the higher speeds that became possible for ploughing as tractor power developed. It tended to break at the tip and a stronger frog known as the Epic took its place. The first of the bodies subsequently used was the Epic 665 introduced in July 1946 which was similar to the IDCP39, with the Epic 39 and 41 introduced in October 1953. This plough was sold both in the home and export markets.

This body was further developed and became the FRDCP Fordson Ransomes Double Chilled Plough which was similar to the Epic 41. This was lighter and designed to suit the mounted ploughs introduced in the early 1950s. It was first introduced in April 1954, and unlike the IRDCP/Epic versions the cutter was incorporated within the breast and was not a separate feature. The number of points available for this plough type comprised only eight or so different patterns.

Digging ploughs

For slightly deeper work and where a more broken surface texture to the soil was required, the first plough introduced for tractor work was the TCP Titt Chilled Plough, a chilled cast iron plough of which the TCP69 was the first variant introduced in December 1920. This body continued in production until the introduction of the mounted ploughs, just after the Second World War. There were two further chilled plough variants, the LCP, the Lincolnshire Chilled Plough, initially manufactured in July 1892, with the LCP21 being the first used for tractor work in February 1935. The SCP, the Steel Chilled Plough, was an earlier introduction in September 1884, the SCP71 being introduced for tractor purposes in May 1930.

The AGT was introduced in January 1938 by George Johnson in Scotland with the adaptation of an export body called the Guidtop plough from South

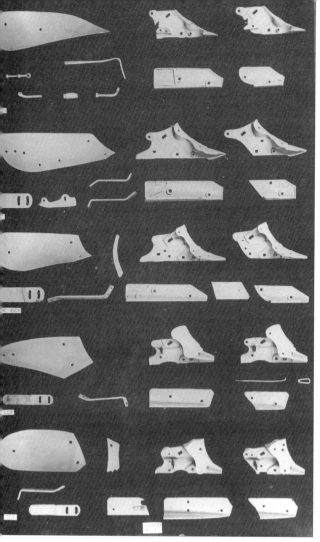

A selection of bodies produced for the home market just after the Second World War. From the top they are: YL, GT, Epic/IRDCP, LCP and TCP.

Africa. The GT19 subsequently became renamed as the AGT3, which was used for ploughing where a complete inversion of the furrow slice was required – a popular feature of Scottish ploughing.

What we would now know as the true digging plough was not introduced for use with tractors until the UN, Unitrac, in 1930, but before that a plough popular in export markets, originally used in South Africa and known as the DM Demon Plough, had been introduced in September 1919. Later versions were introduced, through to the DMC52 which was brought out in 1951, and a further version, known as the DMD, was produced for the TS54 Robin in September 1950. This body was extensively used both at home and abroad.

These ploughs were initially classified as diggers as opposed to the general-purpose and semi-diggers that were in extensive use, but in practical terms their depth and width of cut was less than that of the diggers subsequently introduced and it is interesting to note that Ransomes occasionally classified these ploughs latterly as semi-digger in type.

A version known as the EFR - English Ford Ransomes - was used by Ford as the digger body for the early Ford Ransomes mounted ploughs. It was similar to and copied from the DM plough to the extent that when Ransomes took over manufacturing the Ford Ransomes ploughs in 1955, this plough breast reverted to that designation. The EFR, introduced in February 1947, had a cut of 12 inches and was similar to the DMC3, and another version, the EFR7, had a cut of 14 inches.

With the introduction of the Unitrac plough in 1930, a body of greater depth and cut was made available.

A selection of DM bodies and accessories much used in export markets.

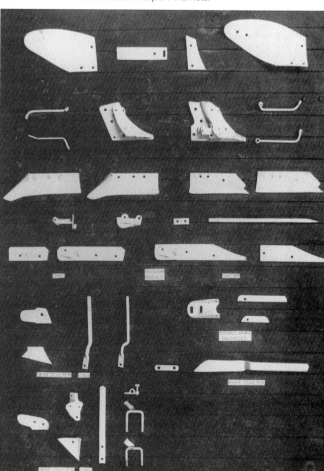

The UN3 was introduced in November 1930, continuing until the development of the UN33 in March 1939. This body was one of the first specifically introduced for tractor work, most of the body types previously described having been developments of ploughs first used for animal-draught purposes.

With the introduction of the Duotrac plough, the Unitrac body designation took the second letter D, the UD3 being initially introduced in August 1931 and continuing until the UD34 in May 1949.

A further variant, known as the UG, Unitrac Guerri, was the derivative of a breast which was brought back from the Italian plough company Martinelli by David Ransome in 1931 and continued in production until 1937.

Further variations continued, with the NUD superseding the UD. This type continued until September 1954 with the introduction of the NUD11, which was a final adaptation of the UN of the 1930s.

A further body type known as the JUM, based on the Jumbotrac plough, which had been developed for the African market, was introduced in October 1910. The first tractor variant, the JUM7A, was introduced in May 1931, with subsequent versions being produced for a number of deep digging ploughs.

Further variation came with the introduction of the SCPT for the Unitrac plough in January 1929, and this continued for some years of production, finally being used on the Twinwaytrac in August 1935.

A number of specialist digger body types were also introduced at this time, including the SL Swampland, introduced initially for export to Estonia in the 1930s. While all these digger bodies were significant for their width of cut and depth of digging, none of them compared with the SU developed for the Supertrac plough in November 1944, with a further variation being introduced in June 1949. This plough, which will be described in detail a little later, was the biggest manufactured by Ransomes and they produced nothing else like it which ploughed at a greater width of cut, let alone with four furrows!

The year 1920 saw the introduction of the BP - Bar Point - body for use with tractor ploughs, which was specifically designed for stony and rocky soils. These types of soil condition led to relatively frequent damage and breakage of shares, leading to greater costs of replacement. The bar point comprised a square section bar forged to a chisel point which projected through a hole within the body of the plough and in effect became the tip of the share. As it wore it could be forged with a new chisel point and was regarded in areas of stony or rocky ground as indispensable. In the 1980s it was described as costing "the price of two shares for the life of six."

The requirements of the bar point body led to another variant, introduced for the Scottish trade, and this was the SCN digging body. It was initially introduced as the SCN3K in May 1951 and used for the TS64 digger plough and the TS46E Multitrac trailing plough. A version without the bar point was

The TCN body took over as a semi-digging body from the Epic/FRDCP ranges in the late 1960s.

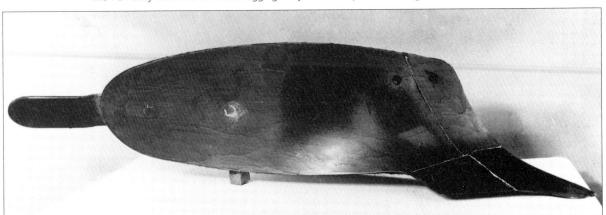

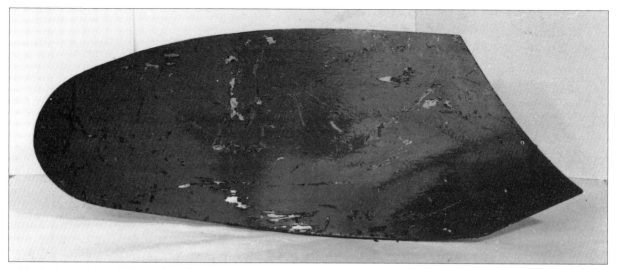

The TCN was quite quickly supplanted in the 1970s by the UCN, which stayed in production until the end of plough manufacture.

introduced in 1960 and first used with the TS82 reversible plough. The SCN introduced an even more radical concept to mouldboard description. This body was based upon a design by a Scottish farmer's son, Jimmy Gass, who was an outside representative for the company and expert in ploughing techniques. That year had seen the sacred Scottish relic called the Stone of Scone stolen from Westminster Abbey, where it was usually kept, and taken to Scotland by a group of Scottish nationalists. Jimmy Gass, being a proud Scot, was much taken by this event, so much so that when it came to classifying the new mouldboard he named it the SCN as an abbreviation of the name Scone!

The CN part of the designation continued with the introduction of TCN3K in February 1964. This ploughed at a shallower depth than the SCN, and was the successor to the semi-digger range supplanting the Epic body types and the DM and FRDCP variants which had been extensively used until then. A further digger body in the form of the UCN, with a 16 inch cut, was introduced in July 1972. This was re-marqued as the KCN in 1977 and again a few months later as the RCN. A further version, the MCN, based upon the same body type, but longer, had been introduced for the World Ploughing Championships in 1971. In 1982 the UCN marque reappeared, this time as a semi-digger body, supplanting the previous variants, manufactured in boron steel, made at the Steel

Case Manufacturing Company in South Wales.

The final range of bodies available for the TSR300 range included the UCN-D, SCN-Y, the YCN, a shallow ploughing replacement for the YL, which dug up to 8 inches deep and cut a furrow up to 14 inches wide and the SLT. This was a slat body for very heavy soil conditions and cut at a similar width and depth to the YCN. The UCN-Y-BP had a sprung bar point to minimise damage to the forged share point.

Special purpose ploughs

Among the more specialised markets for which ploughs had been developed, was that for the horticultural trade. The VY, the Victory, was used for such purposes, being based upon an export plough. It was an orchard plough, which originated in 1923, and made for an early horticultural plough which was developed into the MG2 crawler.

There were further horticultural versions such as the EC Eckert Cape, first introduced for the South African and South American trade in 1903 for ploughs manufactured by Eckert in South Africa.

The SHP Small Holder Plough was based upon the derivative first introduced in 1893 for a company in Maidstone, Kent and the SHPM5 in May 1952 was the version produced for the mounted plough used with the MG6 tractor. One of the YL derivatives was also used for the same plough, the YL167, and an IRDCP body, the No. 7, was also used by the MG tractor ploughs.

Another type, the RHA Ransome Hall & Co., was introduced in 1937 for the TS25 and 30, with a subsequent version being introduced for the TS30 and 31 in April 1938, this being a ridging body, first introduced in 1893 for use with ponies.

Seed covering or riffler ploughs were popular in some markets, where they were used for stubble paring or initial soil preparation prior to deeper ploughing in the winter. These took a number of the special bodies, described elsewhere, of which the VY was an important derivative.

Twentieth century development

The First World War and its aftermath had a significant effect upon Ransomes' trading activities, but rather less upon technical development. The Second World War, however, was different, and trading recovered more quickly. Ransomes' commercial activities became influenced more by production than by sales, and the plethora of plough bodies that were introduced over the years was a consequence of trying to satisfy ever more diverse requirements and markets. At one stage the works had no less than 300 separate dies for use in the manufacture of the many different plough bodies. When one considers that by the time reversible plough ceased development the number had been reduced to a handful of types, this gives some indication of the enormous problems that arose in terms of manufacture in the pre-war era.

After the Second World War, a number of older digger and semi-digger variants such as the LCP, TCP and SCP were discontinued, with the IRDCP and Epic versions continuing. This period also saw the elimination of the Newcastle body, and it is of significance that only two types of mounted plough were able to use this body, the TS63 and TS72. The DM and its variants continued after the war, as did the various Unitrac and associated versions, these having a specialist appeal where soil conditions and root vegetable cultivation necessitated the use of ploughs that could dig at considerable depth and width.

By the end of the 1960s, rationalisation had eliminated all the 19th century plough types except

the YL, and by the late 1970s and early 1980s the only types available were the SCN, TCN, UCN and YL variants, together with one or two specialist types which merged into subsequent variations such as the UCN and the Bar Point types. The SLT was introduced in the same decade. During the 1980s the TCN variant disappeared prior to the introduction of Ransomes' final plough, the TSR300. At the same time the YL, the earliest of the major prize winning ploughs, ceased to be manufactured after 140 years, supplanted by the YCN.

The early tractor ploughs

Having set the scene of the body types, we can now consider the individual ploughs that effected the transfer from the age of the horse to the establishment of the tractor and its trailing plough as the major force in cultivation on the farm.

The development of the internal combustion engine in the late 19th century saw the emergence of a diverse number of applications, for both transport and general power. Agriculture was an obvious field to benefit from this form of engine, and around the turn of the century a number of manufacturers began to develop tractors using the new technology.

Ransomes were quick to exploit the advantages of these early developments, and in 1904 the first evidence of the specific development of a tractor share plough was seen in the model known as the DTP Deep Tractor Plough of single-furrow form. In the same year a three- and four-furrow tractor taking the YL body was developed for use behind Ransomes' own tractor, but it was not a great commercial success and the company abandoned tractor development. It was, however, later to have considerable success with its MG range of horticultural crawlers.

Other manufacturers including Saunderson and Ivel were developing agricultural tractors, and for the Saunderson, Ransomes developed three- and three/four-furrow ploughs, both taking a digger body which had been developed for the export trade, the RSA Ransomes South Africa. In 1909 the first general-purpose plough was produced for the home trade called the YLTM Yorkshire Light Tractor Multiple,

One of the first photographs of a tractor-hauled plough in the Ransomes archive. It was taken at a Mr Francis's farm at Sproughton near Ipswich and shows a three-furrow plough hauled by an early Ransomes motor tractor. Note the similarity to contemporary cars, although of rather more utilitarian appearance. Ransomes quickly discontinued tractor production although later re-entered the market for market garden tractors.

A single-furrow deep tractor plough - DTP - being demonstrated, reputedly to Prince Schönburg.

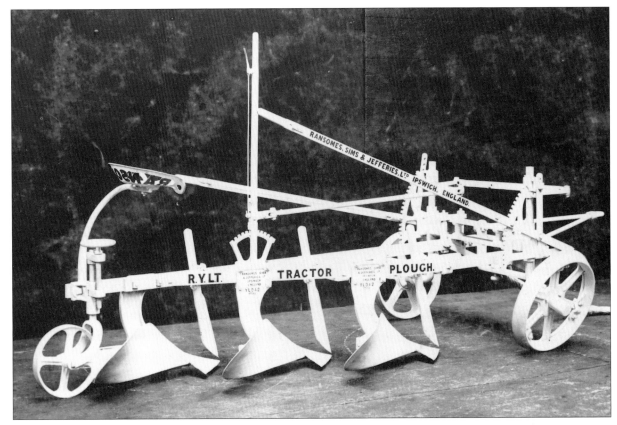

The first mass-produced plough produced by Ransomes was the RYLT and an early production example is shown here in January 1915. It had twin adjustment levers for width control and a single lever for depth which were operated by a man riding on the plough. The plough was attached to a wire rope connected to the tractor drawbar.

with four furrows. Further versions of two, three and four furrows were subsequently introduced, principally for digging where there were specific requirements for the overseas trade.

It was not, however, the multi-cylinder internal combustion engine alone which was being used for agricultural purposes, but also the internal combustion agricultural tractor, developing power from one or two large cylinders. These were more akin to the earlier steam engines, from which they had been developed, and which they looked like in size and appearance. In 1913 Ransomes developed the RMTM Ransomes Motor Tractor Multiple with the RMTD Ransomes Motor Tractor Double, which were introduced as four- and two-furrow ploughs. Most of these were sold into the overseas market, but some were used, to a limited extent, in the home market. All of these ploughs were developed for large scale users

or for large estates where capital was available to meet the substantial costs of the early tractors, and early photographs show some of them used in tandem behind such tractors. The small farmers who dominated the home market were not able to afford such tractors or the ploughs to use behind them.

The year 1914 saw the introduction of the RYLT Ransomes Yorkshire Light Tractor, which was probably the earliest general-purpose plough and which, from photographic evidence, seems to have been in relatively common use, although the numbers involved would probably not have been substantial, given the comparative rarity of tractors. It was designed for early tractors developing relatively low power and available as a three- and four-furrow variant.

It needs to be remembered that all of these early ploughs were what were known as riding ploughs.

Most of the early tractors used behind them would not have run "in the furrow" in the conventional sense, but "on land" and consequently in the early days required a man on the plough to steer and keep the plough straight. It was not until later developments that tractors were able to run "in the furrow" and be operated by one man. The economics of running a tractor-hauled plough with two men meant that, unless three or four furrows were being turned, there was not a great deal of advantage over the horse plough. On lighter soils two-furrow horse ploughs could be operated by a single man, hence four furrows could of course be turned by two. In these early times there seems to have been a certain amount of adverse criticism of the practice of tractor wheels running in the furrow, and it was considered that the ensuing wheel compaction harmed the soil structure and led to potential loss of fertility.

Contemporary with the RYLT was a plough introduced in 1913 for export purposes, of two-furrow form and known as the Emperor. It was a general-purpose plough in the sense that it was available for a wide number of applications, whereas most of those previously introduced had been manufactured, at least initially, for specific soil conditions and markets.

During the First World War, a plough of simple design was produced in two versions for the War Department. This was designated the RTPYL Ransomes Tractor Plough Yorkshire Light and the RTPTCP Ransomes Tractor Plough Titt Chilled Plough. This latter type used an early digger body offered by Ransomes as an alternative to the general-purpose YL pattern where deeper ploughing was required. This plough continued in production, together with others from the pre-war period, throughout the First World War, but it was not until the year after the cessation of fighting that the introduction of new plough types began once more and the first of these was to set the pattern for future trailing plough development.

The RTPYL was specially manufactured for the Government in the First World War, probably for use with the early MOM Fordsons that were being manufactured in the United States in 1917. This plough was first manufactured in the spring of 1918.

CHAPTER 5

The HEYDAY of the TRAILING PLOUGH

The introduction of self lift

The First World War had a very dramatic effect upon most aspects of life in Britain, including agriculture. The immense number of casualties, both dead and wounded, led to serious difficulties in running commerce and industry, and this was compounded by the huge requirement for the wartime use of draught horses. Although internal combustion engine lorries and other motive power were used for transporting goods and munitions to the battle front, horsepower played a very large part in such haulage. There must have been serious difficulties in breeding a sufficient number of horses to satisfy the huge requirements of the war, and there is little doubt that would have been a factor in the Government's urgent search for a cheap and reliable tractor for agricultural purposes.

While a number of national manufacturers attempted to fill the breach, the effort required to produce the munitions and other important items of war meant that the country could not provide the resources in materials and tooling to develop and manufacture such a machine in quantity. The Government looked overseas and found the answer in a tractor which had been developed by Henry Ford and his engineers in North America. That tractor, the Fordson, subsequently became known as the model F, although earlier models were known as the MOM (Ministry of Munitions). It was imported into the UK in large numbers together with ploughs manufactured by North American companies. The end of the war gave Ransomes a potentially large market for the development of a suitable plough that could be used behind the tractor.

In 1919 this led to the introduction of two different types of plough, of which the RSLD and the RSLM became two of the most famous tractor ploughs that Ransomes ever manufactured. The RSLD Ransomes Self Lift Double and RSLM Ransomes Self Lift Multiple were introduced with a self-lift mechanism. This meant that the tractor operator could lift the plough in and out of work by means of a lifting mechanism actuated by a cord attached to a system of levers operating a curved ratchet that engaged with a

In February 1919 the first RSLD YL was photographed. Note the use of levers rather than screws for adjustment for depth and width control.

sprocket on the wheel of the plough. It was lowered on to the sprocket to put the bodies in the ground, and lifted out for transport or manoeuvring on the headlands. Depth and width adjustment were made with levers, as had been the previous standard practice. The construction of the plough utilised flat metal sections, with cast malleable iron skifes on to which the breasts were directly bolted and fitted with landsides to keep the body straight when being pulled through the ground.

The same year the other significant plough, the RCLD and RCLM, was introduced. This was of similar design to the RSLD/M, but a riding plough, and required the services of an operator seated on the plough in the conventional manner, steering and controlling the depth and furrow width with levers. The development of this plough arose from the continuing controversy over the use of tractors running with two of their wheels in the furrow, but it seems likely that it was not continued in production for long before self-lift ploughs with wheeled tractors came to dominate ploughing, except for those where the wheels were too wide for the furrow

bottom and had to be driven "on the land".

A further feature of these ploughs was that they were described as being suitable for low-powered tractors, which is in reality what early wheeled tractors were. The RSLD/M was capable of a considerable amount of adaptation, and a rear lift was made available with a chain-driven pulley, later operated with flat metal bars for easier lifting in and out of the work in order to reduce the width of the headland. The plough could also be used as a single-furrow plough and fitted with a subsoiler in place of one body. Bar point bodies were introduced for the Scottish trade, where this particular type was to become popular. This plough was a great success commercially and won prizes and medals at a number of events, most notably in the 1919 Lincoln Trials, the East Suffolk Show and at a meeting of the Royal Agricultural Society. The plough was available for export with digging bodies, although for markets such as New Zealand, where English style ploughing was popular, YL bodies were made available.

The plough was soon developed into a number of versions and the No. 2 was fitted for the first time with

A strengthened adaptation of the RSLD was introduced in the summer of 1926 and used in connection with the TCP digging body which was popular in the early 20th Century.

An RSLD at work fitted with special attachments for further cultivation on arms at right angles to the main body of the plough. Note the position of the rear wheel. Seen on a farm near Ipswich behind an early International.

screw controls for depth and front-furrow adjustment as opposed to levers. The No. 3 was adapted with the traditional "L" pattern handles which continued in use to the end of production. Version No. 4 was fitted with bar point digging bodies with a more robust "I" section steel frame, offering a greater depth and cut than was available with the earlier variants. A number of subsequent changes were made incorporating further variations, but it was not until the No. 9 that larger scale production began of a general-purpose plough manufactured for light duties with a cut of 8 to 11 inches and a maximum depth of 8 inches. The No.

10 was a further variant for Scotland, and the No. 11 was a four-furrow type developed for easy conditions, but it was not put into production.

The plough continued to be developed until after the Second World War and the Nos. 12 and 13 were introduced in 1946, the latter taking bodies that were popular in Scotland. In 1948 the final version, the No. 15, was produced, taking most of the standard bodies in common use at the time including the YL, TCP, RND, LCP, GT, IRDCPT and Epic, with a bar point variant for the No. 10 and 13 models. Separate adaptations were introduced for the overseas trade,

An RSLD No. 9 with RND bodies awaiting despatch from the town warehouse - Ipswich – in 1937.

An RSLD No. 15 fitted with YL bodies shown in its final development and photographed in the summer of 1947. Compare this with the first production model shown earlier in the chapter.

taking the standard designation of RSLM, generally of three- and four-furrow type, and used with overseas digger bodies, and the four-furrow version was particularly popular in New Zealand.

In 1933 a version of the three-furrow RSLM was introduced for orchard work, designated the TS12, the TS12A being the two-furrow variant. Three years later the RSLD and M Major were introduced, taking YL or IRDCPT bodies, specifically made for use in Essex, with a greater under-beam clearance. Three years later the TS35 was introduced, another cut-down version of the RSLD/M for the hop trade.

The parallel development of the riding plough continued with the introduction of the RST Ransomes Sub-soiler Tractor Plough, taking both the TCP and the SCP bodies, which was a single-furrow plough provided with a sub-soiler, and an export version was also available, taking one of the overseas digger bodies. This was developed as a conventional plough in 1922, taking the same digger bodies either as a pair or as a single-furrow plough with a sub-soiler. Sub-soilers were subsequently offered for a number of trailing plough types for use where farmers wanted to break up compacted soil at a greater depth than the

The RDS had been introduced some years prior to the Second World War and was a popular plough used for semi-digging work. Here we see a late example the RDS No. 4B in the summer of 1947.

bodies could turn. A stronger variant of this same type of plough was introduced in 1927 as the RDS Ransome Digger Scotland, but with the self-lifting mechanism, taking a similar arrangement of digger bodies with or without a sub-soiler. A further multi-furrow plough, the TS38, was introduced for the home market, with a four-furrow version for Australia, taking the TCP and the YL in three- and four-furrow variants. The plough was finally developed as the RDS Nos. 4A and 4B in 1946 in three-furrow form, taking the TCP digger body. This plough was a contemporary of the RSLD/M, but more robustly built and with greater underbody clearance to allow the use of the digger bodies in "dirty" ground.

A revolutionary design

In the second decade of the 20th century, Harry Ferguson began his experiments with mounted ploughs, which led eventually to his patented design for draft control. These experiments utilised a Fordson F tractor and special plough, but had in fact begun rather earlier, with an experimental adaptation of a car called the Eros. Ransomes developed a small two-furrow plough capable of being mounted on a mechanical linkage on the back of the Fordson tractor.

This was known as the Weetrac and introduced commercially in 1927. It was capable of being fitted with both the YL and the LCP bodies and was a general-purpose plough. Early photographs show that it was developed for the Fordson but was also subsequently used with the International Harvester 10/20.

It seems unlikely that this plough was sold to any significant extent and, although Ferguson's experiments with the hydraulic linkage and the mounting of implements had begun ten years previously, it was not until the introduction of the Ferguson Brown tractor in the middle of the 1930s that the concept of the mounted plough came into more general use. A plough had been developed by Ferguson for use with this tractor made by another manufacturer and this exploited his draft control mechanism. It was unlikely that there were sufficient numbers of these tractors sold in the UK market to have made large scale manufacture of this type worthwhile for a company like Ransomes. It was not until the Second World War that the larger scale introduction of mounted ploughs occurred and Ransomes' involvement with these will be dealt with later

The Weetrac was an early example of a competitor to the Ferguson tractor plough. It was mounted on a mechanical linkage rather than one actuated from a transmission-driven hydraulic system. The lift used a substantial spring for compensation purposes but the plough was actually very light in weight, little more than 400 pounds.

A Weetrac at work behind a Fordson F tractor.

The Motrac family

In 1928, the year after the introduction of the Weetrac, came the development of a further plough type, which became known as the Motrac. This was initially developed for small fields and low-powered tractors, and was made to compete with a recently introduced Oliver plough. This was of different design to the RSLD/M types and used forged swan-neck beams instead of the separate skifes, which was a radical departure from usual manufacturing practice. The swan-neck beams were a plain forging to which a separate cast frog was connected, which was made with different shapes for the varying styles of breasts which could be fitted to the plough. This was a different concept in plough design from the practice that had been used for the RSLD and M variants, although later models did take a detachable

frog fitted to legs rather than skifes, which fell into disuse.

The Motrac was developed into a range of differing types of plough, with both fixed and hinged drawbars, and it was capable of being fitted with bodies ploughing with a cut of between 8 to 11 inches and a depth of 3 to 10 inches. Subsequent versions were introduced for use behind tractors ploughing small fields, with even a special type, the No. 4, for

In early 1928 a further general-purpose plough was developed called the Motrac and an example in two-furrow YL form is seen in late May 1928, when first manufactured. Note the absence of skim coulters.

The period after the Second World War saw further manufacture of earlier types of plough and the TS43 Motrac is seen here fitted with YL bodies. The long operating levers make it likely that the plough was intended for use with the early Fordson Major tractor.

ploughing ridge and furrow fields. The plough could also take an optional rear wheel, depending upon its length and the need for turning in shallow headlands, and export versions were made, taking the Demon bodies and fitted with one, two or three furrows. A strengthened variant known as the Major, with a stiffened beam, was introduced in 1936, which was capable of taking a fourth furrow and could be fitted with YL, IRDCP and LCP bodies.

The Motrac continued in production until after the Second World War, and in 1946 it was redesignated the TS43. The frame was manufactured with a proprietary form of steel beam known as Firthag and

its final designation, TS43C, was introduced in December 1950. By then the mounted plough was being introduced in increasingly large numbers, and by the middle fifties the dominance of that type was of such significance that sales for trailing ploughs were in rapid decline.

The Motrac was one of a family of ploughs that were made of similar design, and in 1929 a heavier duty version known as the Midtrac was introduced specifically for the Scottish trade. This was initially available with the IRDCP and LCP bodies, but subsequent versions, introduced in the 1930s, also took the YL, GT, LCP and IRDCP bodies. An export version was developed, known as the TS14, with lower depth screws and a plain drawbar, but this was not put into production. After the Second World War the plough was redesignated the TS45, and

The Motrac was to constitute the first of a range of similar ploughs of which the three-furrow Midtrac was of more robust construction, although later variants of the Motrac could also take three furrows. It is seen here fitted with IRDCP breasts.

The GT body which was popular for use in Scotland is seen here on a No. 4 Midtrac and was built immediately after the Second World War. Note the strengthening to the frame.

continued in production alongside the Motrac for heavier duty work. A further derivative, the TS58 Midtrac Major, was introduced in 1950, designed as a multi-furrow general-purpose plough to be used in light soils.

A further variant, the Multitrac, was first made in 1929. This was capable of being used as a four- and three-furrow plough and initially took the YL and LCP bodies, with later variants introduced in the 1930s also taking the IRDCP type. This plough was of substantial construction, and probably used more frequently behind crawlers than with wheeled tractors, which were inadequate in power output to handle such a substantial plough. Re-designated the TS46 after the Second World War, the Multitrac was one of the last trailing ploughs to continue in production, the final versions being manufactured in the 1960s.

In 1931 a five-furrow version known as the Quintrac was introduced, taking similar bodies to those previously described, and this continued, designated TS7, until the last type, the No. 7, was introduced in 1940. Five years later, a six-furrow version, the Hexatrac No. 1, was made and this again continued in

The TS58 Midtrac Major was regarded as a multi-furrow plough for use on light soils. Here we see an example being tested in work behind one of Ransomes' Fordson Major tractors. Notice the robust construction and hence the plough designation Major.

production for some years, the final variant, the TS69, being introduced in 1952. This was also a heavy duty plough that would have had to be used behind crawlers for normal cultivation.

Ploughs for special purposes

Special trailing ploughs were manufactured for those areas where differing styles of ploughing were in use, and the seed-covering SC was introduced in 1922, taking special "paring" bodies for seed-covering purposes in five- or six-furrow form. These were small bodies which were used in land which did not require much surface cultivation other than general clearance for the planting of seeds. A subsequent variant, known as the Riffler, was introduced in 1927, again a five- or six-furrow form taking an experimental body and later the Victory body.

In 1935 a balance plough known as the Twinwaytrac

An early multi-furrow plough was the PE165, an interesting example of an experimental plough that went into production for an overseas customer. These bodies were used for seedbed preparation rather than complete soil inversion and were popular in some locations where soil did not require extensive surface tillage.

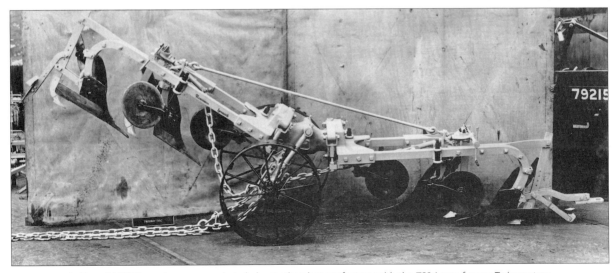

In the mid 1930s a return was made to balance plough manufacture with the TS24 two-furrow Twinwaytrac.
The plough was made for a customer in Lincolnshire but only one batch of five was manufactured.
It is seen fitted with the UD bodies with which it was designed and at the rear of the left-hand beam can be seen a sub-soiler.

TS24 was developed, taking the UD and SCPT body type in two-furrow form. A single-furrow version, the TS33, was introduced at the same time, but adapted for use with the SCPT body only. This was an interesting throwback to the previous popularity of this form of plough for use behind animals, and was designed as a means of trying to overcome the problems of having to plough "one way" and to lessen the costs of such ploughing. The tractor used was a crawler, which turned on the headland, pulling the haulage chain under one of the plough wheels, and continued ploughing by lowering the other set of bodies in the ground and raising the two that had been in use, which were carried raised behind the tractor. Only a small number of these ploughs were made for a special order.

Ploughs for the market garden

In 1919 Ransomes had developed a motor plough which was a dedicated ploughing machine provided with its own fixed internal combustion engine, and fitted with two or three furrows of the YL body type. This was introduced for the home trade, and the following year, 1920, saw a variant with the Ransomes South Africa body - the RSA - produced for the export trade. These ploughs were really only suitable for small-holdings or market gardens and in 1921 Ransomes introduced a further experimental plough for similar use, known as the NB, which was a single-furrow plough taking a horticultural body introduced for the Brittain garden tractor. In 1935 the TS25 was introduced, capable of being provided with one or two bodies of general-purpose or digger pattern, for the MG2 horticultural tractor. This was further developed with the introduction of the TS30, taking a single furrow with a variety of bodies including RHA and EC pattern. The TS31, a two-furrow variant, was introduced the same year, taking smaller bodies such as the VY, SHP and RHA. A replacement plough was introduced in 1945 with the introduction of the TS42, taking the same pattern bodies and in addition the YL and IRDCP type, and a lowered version for orchard work was introduced the following year, designated the TS42A.

In 1951 the MG6 tractor was introduced, capable of being fitted with lift arms of a pattern similar to those

The TS42 was a one- or two-furrow plough manufactured for use behind the Ransomes MG2 horticultural tractor, seen here on test at an unknown location.

The introduction of the horticultural MG5 saw its use with a small form of mechanical lift, on which could be fitted either a TS65 or TS66 for right- and left-hand side ploughing. This tractor thus offered the facility for an early form of reversible ploughing.

introduced for the hydraulic lift on contemporary larger tractors, and two separate ploughs were produced for this linkage, which was operated by a single hydraulic ram. The separate lift arms were fitted with a right- or left-hand body. The TS65 was the right-hand bodied version, the TS66 the left and they were utilised in a similar manner to a reversible plough, with alternate bodies being lifted in and out of the work.

Ploughs for home and export

In the same year as the introduction of the RST, an export plough was made available, following on from the Emperor of 1913. Called the Dictator, this was provided with two and three furrows and fitted with the RSA digging breasts. Renamed the Autocrat and taking similar bodies, this was introduced in 1921 for other markets. A further variant for the export market, known as the Consul, was introduced in 1927 and this was developed up to the Consul No. 4 in 1936, taking the DM bodies and available in four- and three-furrow form. This was a heavy duty plough, built with "I" section frames, lever control drawbar, rear wheel and lever controls. At the same time a further range of similar ploughs was produced for various

The TS65 was developed for the MG horticultural tractor fitted with hydraulic linkage. Both the right-hand TS65 and its twin, the TS66 left-handed version, were separately mounted on the back of the tractor and alternatively put in the ground and lifted out of work to provide the ability to plough "one way".

The Proconsul digger plough, manufactured for export, was introduced in 1932. It is shown here in five-furrow version fitted with DM bodies.

The Motrac was developed as a three-furrow export variant; this Motrac XF is shown fitted with an adjustable rear wheel and drawbar fitted with DM bodies.

export markets, including the Vice Consul in 1927, taking the Prince breasts, with the Ambassador being introduced the following year, capable of being fitted with the YL Motrac pattern bodies and made for New Zealand. The year 1933 saw the introduction of the Proconsul, taking the DM bodies, with bar points if required. It was of heavy duty construction and manufactured as a three/two- or a five/four-furrow version. This was developed until 1939, when the

TS3C Proconsul was introduced in four/three-furrow forms. This plough was intended for less robust work than that for which the Consul had been introduced.

Export versions of the Motrac were separately developed in the thirties, and the earliest model, the TS5, was introduced for the Italian trade and fitted with an experimental body and fixed drawbar. This continued in production for a few years, and a subsequent variant was introduced in 1936, taking the

The Unitrac plough had been introduced in the late 1920s; in 1932 a strengthened variant, the Unitrac Major fitted with UN bodies, was put into manufacture.

range of digger ploughs was manufactured, of which the Unitrac type, manufactured in the late 1920s, subsequently evolved into a diverse range with the ability to take various Unitrac deep digging bodies. Some ploughs took the JUM Jumbo bodies and were of heavier duty application and took the trade name Major, with others for less substantial duties taking the Minor designation. This plough type continued in production through the 1930s and was extensively exported to a number of markets, but also found use on British farms. Some were provided with special wide wheels for use in swamplands overseas, and these large ploughs were capable of turning a furrow of up to 18 inches in width and 16 inches in depth using the SL Swampland body.

The Deeptrac, introduced in 1929, was another type manufactured for export, again taking the JUM and subsequently LCP and DM bodies, and produced in two- and three-furrow versions for strenuous digging conditions. A similar deep digging plough, the Deep Furrotrac, was developed at about the same time.

DM body in a four-furrow form, made for Holland. A twin-axle version was built, and three- and four-furrow variants subsequently introduced for Italy. After the Second World War the plough was re-designated the TS44 and curiously renamed the Litrac. It was based on the design of the TS43, manufactured for the home trade, and this plough continued in development until 1954.

Overseas plough usage in the 1920s appears to have indicated that the original adaptations of the general-purpose ploughs - the RSLD/M pattern - were not robust enough for digger plough use. A separate

A four-furrow export plough, the TS10 based on the TS4 Giantrac, was introduced in 1932 fitted with UN bodies. Ploughs of this size and depth could only have been hauled by crawler tractors, given the power requirements to turn over such large quantities of soil.

When Ransomes introduced the TS (Tractor Share) numbering system, the first plough that was so labelled was the Solotrac introduced in 1931. Manufactured for Italy, it was fitted with the UG body that could turn a furrow 18 inches in depth.

The RAPT plough for the Argentinian trade was also first made in the late 1920s. This was subsequently developed to take between three and five bodies with the Demon digging breasts. This was not in production for long, probably manufactured for a specific overseas agent.

In 1931, further ranges of digging ploughs were introduced, of which the TS1 Solotrac was a single-furrow type, initially taking a special body based upon the Guerri, for the Italian trade. This was made to take the Unitrac bodies with its variants of UN, UD and UG, the latter taking the bar-point body. These ploughs could be adapted to take wide wheels required for support on very soft ground. A number of the Solotracs were also sold at home where Fenland farmers found them ideal for their needs.

In the same year, the two-furrow version known as the Duotrac was first made, taking an adapted UN body, the UD, of smaller size. This could be adapted to take a bar point, and a stronger version, known as the Major, was introduced later that decade. This plough could cut between 12 and 16 inches in width and up to 12 inches in depth, and a sub-soiler adaptation was also available. Another export plough called the Junotrac was introduced in 1933, in two- or three-furrow form, taking similar bodies to the Duotrac but capable of greater width adjustment and wider application.

A further export plough known as the Marquis followed the next year, in three- and four-furrow form for working in medium conditions and turning a furrow up to 12 inches in depth and 14 inches in width, again taking the UN and UD body variants. The four/five-furrow variants, the TS19, was made available at the same time.

Some digging plough types had a curious subsequent development. While the majority were a development of stronger form from the general-purpose ploughs, the Junotrac, in contrast, started as a digging plough and was later developed in the thirties as a three- and two-furrow general-purpose variant for export markets, taking the YL body, presumably for those parts of the Empire where English style ploughing was in general use.

The two/three-furrow TS4 Giantrac was initially made in 1931, taking the UN, UG and UD bodies, but

*Another early experimental plough produced in the mid 1930s was the single-furrow Monotrac,
again probably manufactured for a single overseas customer.*

*The TS41 Supertrac represented the largest multi-furrow trailing plough produced by Ransomes. It was initially manufactured
in a batch of 800 for use in India for land clearance work. It is seen here complete with the special SU bodies specially made for it.
The man standing on the right was actually quite small and enhances the apparent size of the plough!*

built with an exceptionally strong frame, stronger than the Duotracs and designed for deep ploughing at a fixed width in strenuous conditions. This was manufactured with fixed multi-thickness beams and a variety of fitments for working in different conditions. A total of twenty-one versions were introduced to allow ploughing at different widths of cut and depth, and these ploughs represented almost the ultimate in choice of furrow width and depth for the trailing plough. They were also sold in the home market,

especially during the war, when the export markets were closed and when hitherto uncultivated land required clearing for arable or grassland purposes.

The final development of the Giantrac was a special four-furrow plough manufactured for the export trade, known as the TS41 Supertrac and first introduced in 1947, but actually designed and developed towards the end of the war. This was capable of ploughing at a width of cut of 16 inches and up to 18 inches deep. It used special SU Supertrac bodies, and in addition could take the UN and UG. This was continued in development until the early 1950s, and an initial batch of 800 was manufactured for the Indian market and, later, another smaller batch for the North African market. This version was a subsequent development known as the TS41A. Among other changes, it was fitted with pneumatic tyres, but these were quickly cut to pieces in the sharp and stony soils. This plough was the largest trailing plough developed and was by any standard an impressive piece of equipment, with substantial plain section steel frames, deep underbody clearance and very large bodies.

The Magnatrac of 1936 had also been a precursor of the Supertrac, but it was designed for three furrows and taking the standard UN and UD bodies. It was only made in prototype form and could cut a furrow from 15 to 18 inches in width and up to 16 inches deep or at a lesser depth with a 4 inch sub-soiler. Both the Supertrac and Magnatrac ploughs were designed for a 75 hp crawler tractor.

The final years

The post-war development of the trailing plough continued, but with a smaller number of the versions described previously and a reduction in body types, with fewer remaining in production. New types of export trailing plough were developed, with a range of three versions of what were known as the Duratrac, of which the TS47 was a three-furrow version, the TS48 a three/four and the TS49 a five/four variant. These represented the final development of the Proconsul type. These were manufactured for the overseas trade, taking the DM and other digger pattern bodies which were popular at the time.

By the mid 1950s trailing plough usage was in terminal decline and the final development of its type came in 1964 with the introduction of the TS85, which was a four-furrow variant based on the TS46D

Some idea of the size of the TS41A can be gained from this view of the rear taken on the quayside at the back of the plough works. The rear drawbar was not a permanent fixture and was used to manoeuvre the plough around the works.

Multitrac, which was made for the home trade. This was fitted with hydraulic operation of the rear lift wheel, and manufactured with wide tyres for Holland, where ground conditions were soft and required the use of special wheels.

The mounted plough effectively ousted the trailing plough for most purposes in the sixties and, apart from the use of five- or six-furrow versions on large arable acreages with crawlers, there was little demand for trailing ploughs from the majority of farmers, who could plough up to four furrows with mounted ploughs. Additional furrows were available in semi-mounted form with the introduction of that type of plough in the early 1960s. The development of the

The TS85, the final development of the trailing plough, was a modified Multitrac. A batch was produced for Holland, partially actuated with hydraulics not seen fitted in this photograph.

hydraulic tool carrier for use behind crawlers allowed them to operate mounted implements, including ploughs, and the rapid development of four-wheel drive tractors with high power outputs gave farmers the capacity to plough with mounted multi-furrow ploughs with what had hitherto only been possible with the largest of the trailing type.

It had taken barely forty years for the tractor-hauled trailing plough to displace the animal draught plough, previously used for centuries, and scarcely thirty years more for the trailing plough to be superseded by the mounted type.

The Hexatrac, one of the last of the trailing ploughs to continue in production, was used behind crawlers and is seen here at the rear of a Fiat.

CHAPTER 6

The FORD RANSOMES ERA

In the early summer of 1945 the Ford Motor Company introduced a replacement tractor for its very successful Fordson Standard. This latter tractor had been a direct descendant of Ford's first mass produced tractor, the model F, and although developed with increased power, by the middle of the 1940s it was technologically behind the competition.

The model that replaced it was known as the Fordson Major, and although it was larger it retained the original engine and gear box of the Standard, providing power to a newly designed rear transmission. The intention was to produce a tractor capable of drawing a three-furrow plough rather than the two regarded as the normal practical limit of its predecessor.

Ford announced at the same time that it was going to develop a range of implements to be used with the tractor. This had been standard practice with the large tractor manufacturers for many years in North America, but was not a practice adopted in Britain. Apart from the obvious marketing advantages and the opportunity to make profit on the sale of implements, Ford knew that competition of a potentially serious nature was on the horizon in the form of the Ferguson tractor, which was to be built by the Standard Motor Company and introduced in 1946.

Harry Ferguson's manufacturing arrangement with Ford in the United States confirmed the view of the North American company about the considerable advantages of this technologically revolutionary tractor. It was a machine designed to be at the heart of a comprehensive mechanised system allowing the tractor to operate a wide range of implements, tools and attachments, thus transforming it from a device that merely pulled, to one that provided a mobile power base for a mechanised farming system.

One of the patents Harry Ferguson had secured was that relating to the draft control of the hydraulic implement mounting system. This could not be used by other manufacturers except, presumably, upon the payment at huge cost for licensing rights, and this was subsequently to have a significant effect upon the design of ploughs available for attachment to the Fordson Major and other manufacturers' tractors.

In 1940 Ford had purchased the Imperial Foundry at Leamington Spa from the Flavel Company, which had been using it to manufacture gas appliances. Ford used the plant initially for the production of items for war work, but when this requirement ceased in 1945, they decided to utilise the factory for the production of a range of implements to complement the Fordson Major.

The first item they produced was the Fordson Elite plough. This was of Ford design and manufacture, based at least in part, on an Oliver design and using bodies based on Ransomes' patterns. It was, however, similar in design to contemporary Ransomes ploughs, and was available in either two- or three-furrow form, but in practical terms, the Major had difficulty in pulling three furrows, particularly in heavy land, despite the design being nominally for that usage.

John Foxwell had started with Ford as a scholarship student at Dagenham in 1936, and in 1939 entered the drawing office at Dagenham. In 1944 he transferred to the tractor section to work on designs for ploughs to be used with the proposed Fordson Major. Ford wanted to use some of Ransomes' bodies, but was surprised to find that they had no drawings available. Ford decided to measure a production item and, using a 2 inch square grid on the side, plan and end, took measurements of the curved shape to make drawings of the mouldboards. The same process was undertaken with the frogs - called saddles by Ford - and shares, and further detailed design took place for manufacturing purposes. That design and development work was transferred to Leamington after the war.

This is the first plough coming off the Leamington production line, a two-furrow Elite, manufactured in August 1945.

The parts that were manufactured subsequently for the Elite appear to have been entirely resourced by Ford, using in-house castings, forgings and machining facilities and procuring other parts from outside suppliers.

From Ford's point of view, an association with a large implement manufacturer such as Ransomes would have been essential in order to produce and market implements of the sort they had in mind within a short timescale. Ford also had arrangements with other implement manufacturers to supply items for the Major. However, these arrangements differed from that with Ransomes because for the most part the implements marketed were standard items from other companies' ranges which were adapted to fit the Fordson tractor but could be equally easily used by competitors' tractors. The association with the country's leading plough manufacturer would give confidence to farmers that Ford's ploughs were reputable, particularly because Ford had not previously produced implements.

Ford's plough manufacturing

In 1946 the arrangements with Ransomes became more formal, and an agreement was made between the two companies for the joint manufacture and marketing of a range of ploughs and other implements. This had significant advantages for both parties. Ford gained design expertise in the earth-moving parts, at the same time utilising existing resources at Leamington for some component manufacture and the assembly of ploughs and other implements. The implements were sold through the existing Ford dealerships, of which there were a considerable number engaged in selling cars and commercial vehicles as well as tractors and now agricultural implements. Ford had previously experimentally produced new types of steel, some of which were being used for the first time in plough manufacture and were of potential interest to Ransomes.

One curious fact remains in that, despite exhaustive research and questioning of individuals who were involved with both companies at the time, there is no information as to the specific nature of the agreement and there is no document in the extensive archive at Reading University! Some have ventured to suggest that this was similar to the famous "handshake" agreement between Henry Ford and Harry Ferguson.

Ransomes gained considerably from the arrangement as well. Ford was the dominant tractor manufacturer in the UK having sold 136,811 tractors between September 1939 and April 1945, and the facility for providing a dedicated range of implements for the new tractor gave Ransomes direct access to a potentially huge market. The ability of Ford to manufacture on a substantial scale was something which the existing resources at the Orwell works would probably not have been capable of at that time. The jointly produced implements were also sold through Ransomes' existing dealerships, increasing further the potential sales outlets. Ransomes would inevitably have been concerned about the potential effect upon their business of Ford manufacturing ploughs specifically for their tractors, because large numbers of Motracs and RSLD/Ms had been sold to farmers with Fordson Standards and later the E27N Majors. The loss of some of this business through Ford manufacturing their own ploughs for the new Major would have been a serious blow to Ransomes' home sales.

Ransomes were still heavily involved in the manufacture and sale of their existing ranges of ploughs, not just for the home market but also for overseas, and with the end of the war the need to produce more food at home much increased the potential market for their existing range of products. Like Ford, Ransomes had been involved in war production and its cessation meant that it was possible to resume normal product manufacture on a larger scale.

The approach to manufacture, and in particular assembly, was very different in the two companies. Ford was dominant in the field of large scale mass production, endlessly refining techniques to improve production, simplify design, reduce costs and increase output. The manufacture of ploughs at Leamington was on a mass production basis, almost certainly the first instance of such in the UK. In contrast, Ransomes were building a considerable number of differing types of ploughs in small areas of their factory. Items were brought from different parts of the works and assembled with the use of simple stands and other such facilities. Some ploughs were assembled by hand and manually moved along tracks for larger scale production.

The average time taken by Ford to construct a three-furrow mounted plough was approximately twenty minutes, but they could be assembled in as little as ten. Pneumatic tooling was extensively used, in addition to mechanised lifting and assembly techniques, and this all greatly assisted assembly. One suspects that the unit production costs for Ford were considerably lower than for Ransomes. Ford used to exhibit ploughs at agricultural shows and hold competitions to see how quickly they could be assembled and taken apart.

Ford commenced the building of ploughs on jigs located on benches, then moved the partially assembled ploughs with electric hoists on to powered assembly lines where coulter arms, disc coulters, skims, wheels and handles were fitted as the ploughs moved along the track. At the end of the line they were removed, spray painted and baked in an oven, from where the ploughs were moved to dispatch. Both assembled ploughs and what were known as "knock downs" were manufactured, the latter being packed in boxes for export. The ploughs and boxes were removed from the factory premises by train and taken to the various suppliers or, in the case of exports, to dockside locations. Much of the information regarding assembly has come from Jack Cox, who in 1946 went to Leamington and subsequently became a charge hand on the plough assembly line.

Bill Lees is another former Leamington employee who worked in the toolroom on machining and manufacture of components for the ploughs. He recalls producing a batch of finely machined pins for securing the shares to the frog. Ford utilised the services of a local farmer who used to make occasional visits to the plant and also experimentally tested

Using their traditional skills in manufacturing on a large scale, Ford built plough frames on fixed jigs among other devices. After partial assembly they were lifted on to an assembly track which can be seen between the two operatives. The plough is a two-furrow Elite.

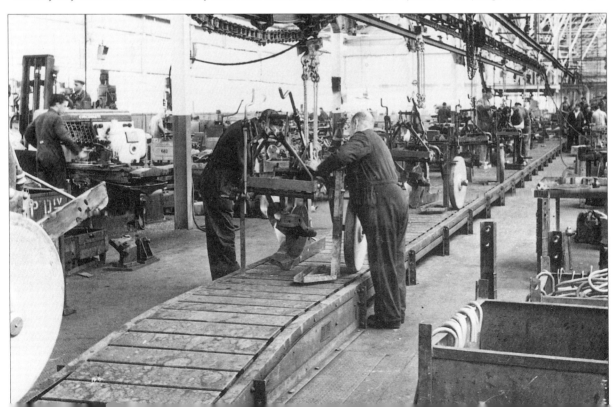

implements on his farm near the works. On one occasion Bill Lees recalled the farmer visiting in a most agitated state: a share had broken while he was testing a plough and he could not remove the machined pin to replace it. Traditionally, wooden pegs had been used for the purpose, and farmers often cut them from a tree in the hedgerow; these were simple to remove if a share broke, but an accurately machined pin not so, given the sort of tools normally available in the tractor tool box! Thereafter Ford returned to more traditional methods of securing shares.

The numbers of implements produced by Ford were substantial and included beet lifters, toolbars and disc ploughs. Over a hundred thousand were manufactured between September 1945, when the Elite plough began production, and the first month of 1952, and it is probable that a substantial number of these were ploughs. This was probably in excess of the numbers of individual ploughs produced by Ransomes, who were producing a much wider range.

In the mid 1950s Ford expanded their foundry facilities at Leamington and required space used by the implement production facilities at the plant. The final plough rolled off the production line on 13 December 1954. In total 84,306 ploughs, of all types, had been manufactured since September 1945. At the beginning of January 1955 Ransomes opened their new plough works at Nacton, on the outskirts of Ipswich, where they had built a new foundry in 1948. Jack Cox recalls that employees from Ransomes were taken to Leamington for training in assembly techniques used by Ford, and apparently expressed reservations that Ransomes would be capable of producing ploughs in the sort of timescale achieved by Ford. Subsequently, all Ford Ransome implements were produced at Nacton, but it seems probable that there was some initial lack of continuity of production, because it is alleged that after Ford gave up production there were delays in meeting the requirements for plough sales, which until then had

A later view of the Nacton production line showing mounted ploughs being assembled. Extensive lifting facilities were available to avoid dangerous manhandling of the cumbersome implements.

been available within two or three days of a farmer's order. Ransomes almost certainly became capable of producing ploughs in as large a quantity and within a similar time scale as achieved by Ford, and the opening of the new works coincided with a reduction in the number of individual types of plough which they had previously manufactured.

The agreement with Ransomes in 1946 coincided with further development of the Fordson Major tractor. The company introduced a rear-mounted hydraulic power lift, which controlled lift arms to move an implement in and out of the ground. This lacked the sophistication of the depth control of the Ferguson system, and consequently a depth wheel was required for adjusting the depth of tillage implements.

The FR mounted plough

Ted Skinner went to work for Ford at Dagenham before their acquisition of the Imperial Foundry at Leamington in 1940. He subsequently went to Leamington, where from 1945 to 1948 he was head of the foundry's Experimental & Development Department and was actively involved in the development of the implements, including the ploughs.

He recalls that the factory received some mounted ploughs in crates which appeared to have been immersed in seawater for some time. These ploughs were used as the basis of the design of a mounted plough subsequently known as the PM — P for plough, M for mounted. The condition of the ploughs seemed to indicate that they had been dropped in the sea, which suggested they had been imported from overseas, probably from North America, and they may have been Sherman ploughs for the Ford 9N tractor which used the Ferguson hydraulic system. Many of the features of the Ford Ransomes PM plough were more similar to those of the Sherman plough than the standard components which were being produced by Ransomes at the time.

Ransomes' involvement in this plough was confined to the bodies, coulters and skims, although there were some detailed variations produced by

Ford demonstrated their mounted FR PM plough on a farm at Rainham, Essex, on 9 October 1946. The demonstration was attended by senior staff from both companies' sales and production departments, including directors. A group of Ransomes' expert ploughmen were also in attendance. Also seen is a three-furrow Motrac and a Ransomes C57 mounted cultivator developed for the E27N Major.

Ford, and again, as with the Elite, nearly all of these components were resourced by the company from their own and outside suppliers, and components were not provided for initial manufacture by Ransomes. Ransomes did produce a number of their standard manufactured parts as spares or replacements, because Ransomes were selling this plough themselves through their own dealership arrangements, and the plough was capable of being used with a number of tractors other than Fordsons. Research through Ransomes' plough components Marks book shows they also produced some parts for the trailing Elite plough, although nominally Ford made all these parts themselves. When Leamington ceased plough production, Ransomes produced wearing parts as spares for these ploughs.

The mounted plough was tested in prototype form in early 1946, and Ted Skinner recalls an interesting incident which occurred when testing was in progress on a farm near Leamington. He was driving a Major and could see that in the distance a limousine was parked at the edge of the field and a small tractor ploughing was being supervised by two well-dressed individuals. One was Harry Ferguson, the other John Black of the Standard Motor Company, which was shortly to begin manufacture of the Ferguson TE 20 tractor, and in fact it was such a tractor being tested. After a while the limousine disappeared with its passengers, but subsequently Harry Ferguson came to the field where Ford was testing their mounted plough on the Major and was surprised to see them testing such an implement. He was, no doubt, concerned to ensure that Ford had not infringed any of his patents on the tractor, and the use of the depth wheel would, of course, have indicated that his patented draft control arrangements were not fitted. It seems a remarkable coincidence that these two important tractors and ploughs were being tested a short distance apart and at the same time, because both were to dominate British agriculture for some years into the future.

On the 9th October 1946 the early production mounted plough was demonstrated to both Ransomes' and Ford's senior staff on a farm near Rainham in Essex. It was preceded on the previous evening by a dinner at the Queens Hotel, Westcliff on

A group of the companies' experts discussing the new plough. Second from the right is Harry Power, Ford's works manager from Leamington, fourth from the right is Reuben Hunt, later knighted and Ransomes' chairman. Fifth from right is the youthful Jim Byewater of Ford, in charge of implement design, and to his right Mick Ronayne of Ford, in charge of tractor development and responsible for the E27N Major. Second from left in the light trilby is Henry Deck, Ransomes' Home Sales Director. The contrast between the youthful Ford staff and the older Ransomes staff should be noted!

Sea, where the two companies celebrated the initial development of the plough. Ransomes also demonstrated a prototype mounted plough at the same event, based on the No. 3a Motrac, although the plough was not subsequently developed commercially. It has to be said that Ford had undertaken by far the greater part of the development work at that time. It is also interesting to note that in inter-company correspondence within Ransomes, the mounted plough is frequently referred to as "The Fordson Plough" and although Ransomes' literature proclaimed it as a joint Ford Ransomes product, there is little doubt that it was regarded by Ransomes as more of a Ford venture. Curiously, Ford's own publicity material gives the impression that the plough was developed and jointly made with Ransomes!

The PM plough, as it was classified by Ford, was made available initially with semi-digger bodies of the Epic type. It was subsequently provided with the popular YL bodies and, in 1947, a new type of digging body was made available. This was known as the EFR - English Ford Ransomes - based upon Ransomes' DM body, thus offering flexibility to farmers to use

Ransomes demonstrated a mounted No. 3A Motrac at the same event. The tripod headstock and other details bear a similarity to the Sherman mounted plough for the Ford Ferguson in the USA. Note the very long top link and the modifications to the tractor linkage. The ploughman was H. Jarvis, who passed away in 2001 in his early 90s.

Three of Ransomes' senior sales staff at the demonstration. On the left is Henry Deck, then well into his 60s and Home Sales Director. On his left is W.D. - Bill - Akester, who succeeded him in that post a few years later. The third of the trio was Fred Ayers, who was East Anglian representative and had been involved in the demonstration and sales of the first mass-produced tractor plough, the RYLT, before the First World War.

The first plough specifically manufactured under the Ford Ransomes agreement was the FR PM, and the number of suited onlookers suggests this is possibly the first to be made.

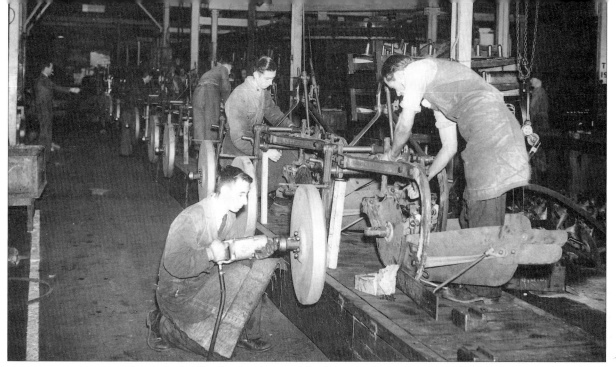

The production line showing the assembly of a range of FR PM mounted ploughs.
Note the powered track and use of pneumatic equipment to aid manufacture.

A completed FR PM plough just out of the paint shop being finally prepared and inspected by a group of onlookers
including, second from the left, Sir Anthony Eden, the local MP, later to become Prime Minister.

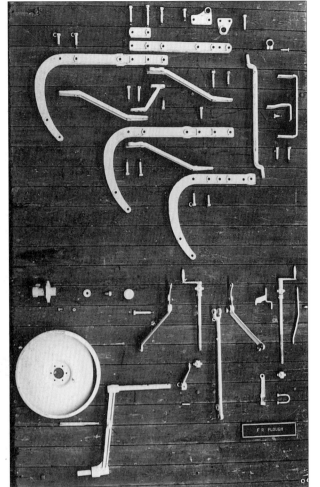

The FR PM mounted plough was not made at Ipswich, but because it was available through Ransomes' agents, they had to be able to provide spare parts. Ransomes produced a standard handbook for the plough and they obtained a set of plough parts from Ford in order to photograph them for the manual.

bodies of the sort appropriate to the soil conditions. This plough was redesignated the EPD and EPE using the number 2 or 3 in the description depending upon whether it was of two- or three-furrow form when fitted with digger bodies. This EP plough was provided with wider cross-beams to facilitate ploughing at a greater width than the PM on account of the use of the digger bodies.

The PM plough consisted of a series of "I" section swan-neck beams, the frame and legs of which were forged in one piece of high tensile steel and interconnected with spacers and stiffening bars of similar form. It was provided with a pressed steel disc wheel for depth control purposes, and screw handles were fitted for both adjusting the depth wheel and controlling the front-furrow width of the plough in work. It was painted in a colour previously called Empire Blue, which was universally used for all of the Ford-made Ford Ransomes implements.

In the summer of 1947 Ransomes started to develop another mounted plough of their own. This was an experimental plough known as PE846 (Plough Experimental), produced in both two- and three-furrow form and designed to plough at a variety of widths. Initially it was fitted with Epic bodies, although it was probably intended to take the YL and DM digger pattern. This plough was developed by Ransomes for its own sales, and consisted of plain section frames to which were bolted swan-neck forged plain section legs taking standard bottoms and breasts. Photographs of the plough show it being mounted

Ransomes appear to have been involved in some testing of the FR PM plough and one is seen at work behind an early Major owned by Ransomes. The company had a prototype of the same tractor in 1945 for implement development work.

This RYLT is shown in original livery, with a colour sometimes called Orwell Blue. It is lighter in tone than the post-war paint.

Detail of the RYLT showing lining out at the rear of the mouldboard.

Further detail of the same plough showing standard early features including the use of skifes for mounting the mouldboard.

An autumn ploughing match with an RSLD No. 15 in use behind an International tractor.

The TS42A was built for the horticultural tractor, seen here in single-furrow form with cut-down controls for use in hop fields, etc.

The TS46 Multitrac was a substantial and robust plough, seen here with cross-beam strengthening made for, and here being used behind, a crawler.

A Leamington-built FR two furrow EPJ fitted with YL bodies awaiting use in a future ploughing match.

Sprung-loaded trip legs were an asset where there was a serious risk of damage on stony ground, so the TSR300T was popular in Scotland and other locations where there was a preponderance of this type of soil condition.

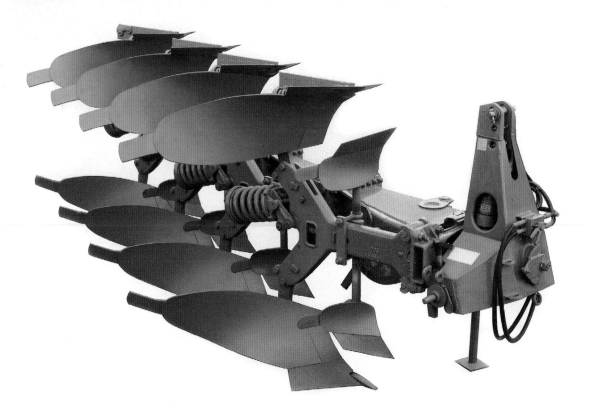

In August 1947 Ransomes developed their own mounted plough, experimentally known as the PE846 and adapted to be used in two- or three-furrow form. Here we see a two-furrow variant mounted on the back of a Major. It was another Ford tractor, the 8N, for which the plough was eventually made when it was produced with some detailed changes as the TS54 Robin.

behind a Fordson Major, although the PM plough was the model specifically produced for that tractor. It was another Ford tractor, however, for which this plough was eventually manufactured and that was the 8N AN produced in the United States and sold in a number of Ransomes' export markets.

The plough took the mark TS54 and was produced in several versions, initially without a depth wheel. In export form it took the name Robin with which it is also known in Britain. In British TS54 form, it took Epic or IRDCP bodies, but subsequently YL and digger bodies. One version, the TS54B, was produced with a depth wheel, initially for use with the Allis-Chalmers B, and taking the mark TS54E, was also made of more robust form for use with tractors fitted with category 2 linkage.

The TS54 Robin was the first mounted plough produced since the Weetrac and is seen here on the tractor for which it was initially developed, the Ford 8N AN. The tractor was provided with the Ferguson patent draft control system and consequently did not require a depth wheel and only had cross-shaft adjustment.

The FR mounted reversibles

In late 1949 a further radical development occurred with the introduction of the mounted reversible plough. Reversible ploughs were not a new concept, having been available since the early years of the twentieth century when they were, of course, hauled by horses. The obvious advantage of a plough which can continuously pass up and down a field without the effort required in setting out the field for one-way ploughing had been obvious for a considerable period of time. Ransomes began to develop such a plough for tractor use towards the end of the 1940s, and in 1950 two ploughs, the TS50 and TS51, were introduced for the Major.

The TS50 was a two-furrow plough taking the Epic body and the general-purpose YL, the TS51 being a single-furrow variant and capable of being fitted with the UD digging body. These ploughs were provided with a mechanical turnover mechanism of some complexity, and fitted with a stay bolted to the tractor axle for stability. There was a very substantial difference in price between the cost of these ploughs and the conventional two- and three-furrow mounted PM series, and it is probably that factor, together with the newness of ploughing with a different technique, that led to relatively slow sales. These ploughs were manufactured utilising special steels supplied by Ford. It was assembled at the Orwell Works and was available under the Ford Ransomes agreement to UK dealers, being sold directly by Ransomes to their export markets. The plough was fitted with small depth wheels and two, of course, had to be fitted, and there was an option of pneumatic tyred wheels in lieu of the steel spoke wheels which were the normal pattern used by Ransomes.

One of the early mounted ploughs produced was the TS50 and 51, the two- and one-furrow reversibles.
Here is the single furrow fitted with deep digging bodies.

To give an indication of the difference between plough prices in the early 1950s, the following is of interest. All the prices quoted are for 2-furrow ploughs:

TS50 £110, PM £50, TS59 £58, TS54A £50, TS63 £68, TS64 £72, RSLD £81-50.

FR plough development in the 1950s

In 1950 a new two- and three-furrow mounted plough was developed for use with the replacement Fordson Major. The new tractor was not to be put into mass production until the end of 1951, but it had been hoped it would be produced earlier, and consequently plough development had begun some time previously. It seems likely that the plough was jointly developed by Ransomes and Ford. The photographic archive of Ransomes shows the

The successor to the PM series, the EPG and EPJ, was manufactured for the new Fordson Major and available in early 1952, and here is the first plough completed. Production of the older PM model continued for a short time and one was photographed only two days afterwards as the 100,000th implement manufactured at the Imperial Foundry since August 1945.

Ransomes undertook some trials with the Ford-manufactured EPG and EPJ ploughs. It appears they made some small modifications of their own with their pattern of skim and arm being used. This plough was actually described as being an experimental model and was probably tested by Ransomes in advance of mass production, hence its being used behind an early and not a later Major.

prototype of what has become known as the EPG and the EPJ series under test near Ipswich, and given an experimental number. This number does not, however, fall within the sequence of those that were being used for experimental ploughs for Ransomes' own purposes, which may perhaps indicate that joint development of the plough range with Ford was regarded as a separate activity by Ransomes. The plough was to be built at Leamington and was manufactured with components resourced by Ford themselves and from outside manufacturers. Parts for the ploughs were stamped with the letters "EP", identifying them as being of Ford rather than Ransomes manufacture. Ransomes, however, did produce parts themselves from their standard range and consequently, as with the PM , it is possible to see both types of components fitted.

This plough differed from the construction of its predecessor. It consisted of forged high tensile plain section beams connected to a front pentagonal shaped casting, to which the beams and headstock connections were bolted. There was a cast steel channel connected to the centre of the frame for stiffening purposes. The malleable iron legs of the plough were detachable, being bolted to the frame, quite unlike the construction of the PM plough which had consisted of a number of slender high tensile steel

swan-neck beams provided with stiffeners and bracings of similar form.

The principal advantage of this plough was that it was probably stronger than the old PM, whose slender legs would not take the same sort of punishment if the plough hit obstructions in the ground, because the legs were part of the frame component. In the event of damage to the new EP model, the legs could be replaced at less cost and damage to the frame, although internal Ford correspondence indicates that these legs suffered early damage, so much so that stronger legs had to be manufactured. The other advantage was that the alteration of the legs to allow the plough to be set at different centres was much more easily achieved than with the PM, where the leg and frame were a one-piece forging. The plough was marketed on the same basis as before, with Ford selling through their own dealership arrangements and Ransomes through theirs. As with the PM, both companies produced their own supporting parts and technical literature.

At the same time Ransomes had produced a similar plough, but of slightly different design, which became known as the TS59 and was developed towards the end of 1950. This plough was probably varied in design for ease of manufacture by Ransomes, because it lacked the front steel casting which was a feature of

In 1951 Ransomes introduced the TS59, which was a three/two-furrow mounted plough for medium-sized tractors. In its early form it was manufactured for the overseas market and here we see one of the originals fitted with DM bodies. Ransomes used standard spoked wheels for a number of years but subsequently adopted the same pressed steel wheel used by Ford.

with both PM and EP types, and eventually the TS59 was adapted to take other bodies developed by Ransomes. Although the plough was marketed as a three- and two-furrow, it was sold in three-furrow form and the handbook contained instructions how to adapt it for two-furrow use. The handbook also indicated that the ploughs would have to be partially assembled by the farmer, but in fact, except to export markets, they were sold in assembled form.

This plough was manufactured in large quantities, and continued in production at Nacton from 1955

the EP and used a standard form of detachable leg of a sort which had been used by Ransomes for many years. This almost certainly made the plough cheaper to produce from Ransomes' point of view. It would appear that it was initially manufactured for export purposes taking the DMD, YL and Epic bodies. It also utilised a standard Ransomes spoked wheel rather than the pressed disc used by Ford. It seems likely that it did not sell initially in large quantities other than through Ransomes' export markets.

Of the four first variants, A B C and D, only one type was put into manufacture; the other three soon becoming obsolete. Thus it was that the TS59E, with category 2 linkage, and the TS59F, with category 1 linkage, became the first mass produced versions of this general-purpose plough and took over

Another early example of the TS59, this time fitted with YL bodies.

from the EPG and EPJ when Ford ceased production at Leamington at the end of 1954.

This plough was, of course, "pure" Ransomes, with all of the detailed design features which were in use on other models. The most fundamental departure from Ransomes' style was the use of the Sankey-manufactured pressed steel depth wheel, which was the same pattern used previously by Ford, although the company had used fabricated disc wheels for some types previously. Bodies were interchangable

until 1967. No less than thirty variations of this plough were produced, and it must have been one of the biggest selling ploughs, if not the biggest, produced by the company. Although it was described as a general-purpose plough, it was capable of being fitted with a wide variety of bodies and consequently used in the widest possible range of soil conditions. There was even a special version produced by Stormont Engineering of Tunbridge Wells, Kent, with cut-down control handles for use in hop fields.

In the early fifties, experiments had begun to take place to add additional furrows to the three-furrow ploughs. Both the early PM and the TS59 had been experimentally fitted with four-furrow bodies, and the older PM plough with a five-furrow body, again as an experiment. This must have been a very considerable weight to mount on the back of tractors at that time, and was probably beyond the capacity, for all practical purposes, of the power of the tractors, and in particular, of the safe use of the hydraulic systems. Mounting such a large plough on the back of the tractor on a hilly terrain would have been potentially dangerous, and the strain imposed upon the hydraulic pump and the lack of ability to weight the front end easily must have been factors which led Ransomes to discontinue such experiments until tractors became more powerful and larger. It was not until 1962, and the introduction of the TS59M model, that the plough was provided with an additional body, and in 1965 the TS59S represented the final development of this model with the introduction of a five-furrow variant.

The extra bodies led to the introduction of the TS59P and R, both of which were fitted with double tubular stays for additional strengthening. By then, the semi-digger Epic bodies were being substituted by the multi-purpose TCN body, with the addition of the YL

The Riffler plough continued to be popular and a version of the TS59 was built with this sub-frame mounted beneath the main frame to take the seed paring bodies.

A three-furrow TS59 at work behind a Fordson Major.

The TS59 plough was adapted for use in a less robust form for the Dexta range and here a two-furrow TS59Z is seen at work.

being provided in areas where shallow ploughing was required.

In 1961 came a further variant of the plough known as the TS59Z, and subsequently the TS59Y, which was introduced for the Super Dexta. Although of similar design, it was provided with two or three furrows only, depending upon whether digger, semi-digger or YL bodies were being used and made with category 1 linkage. The FRDCP body was used instead of the Epic and IRDCP variants, and this was a more compact and lighter component, an important consideration for attachment to the smaller tractors.

Similar to its trailing predecessor, the Motrac, the TS59 was one of a "family" of ploughs which was jointly marketed by Ford and Ransomes, and in 1952 three additional ploughs were introduced for special purposes.

The TS55 was developed as a single-furrow deep digging plough, which took the UDM body and was also capable of being fitted with a bar point. It had been announced in 1951 for the earlier Major but was not put into larger scale production until probably 1953. This body was capable of digging up to 12 inches in depth with a cut of up to 16 inches. The requirement to turn over a substantial volume of soil at this depth and width meant that the available engine power produced by contemporary tractors was probably insufficient to plough anything more than a single furrow. A similar plough had been built by Ford at Leamington.

The TS53 was the earliest of the single-furrow deep digging mounted ploughs. It was not in production long, however, before giving way to the similar TS55.

A two-furrow general-purpose plough was developed in 1951, the TS63. This had a longer clearance of 30 inches between the legs, which allowed it to deal with soil and trash conditions for which TS59 was not suited. This plough was capable of being fitted with the YL and Epic bodies and the

A two-furrow long clearance plough, the TS63, was introduced by Ransomes in late 1951 and an early example is seen fitted with Epic bodies. The plough has been provided with lining out, probably for exhibition purposes, which can just be seen on the beam and stay.

The TS63 was one of only two mounted ploughs that could take the Newcastle bodies, seen here fitted with RNDs.

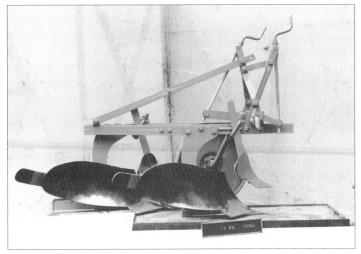

To accompany the TS63, another special two-furrow mounted plough was made, the TS64. This was constructed with bodies with closer clearance for deeper digging and is seen here fitted with the SCN bar point bodies, one of the two earliest ploughs fitted with this new digger body.

Newcastle body - the RND - and the AGT, both of which were still popular in some agricultural districts, and this further extended the potential market for mounted plough sales.

This plough, like the others, was produced by Ransomes with the probable intention of export sales, but it seems unlikely that any occurred and, if so, to any significant extent. This plough continued to be manufactured until the early 1960s.

A further variant, the TS64, was developed the same year, again in two-furrow form but provided with a greater under-beam clearance of 24 inches as opposed to 22 inches on the TS59. The legs on this plough were set closer together than the TS59 and it was only capable of being fitted with Epic and UDM deep digging bodies, and later the SCN body. It was designed for heavy ploughing conditions where there was a particular requirement for a deeper digging plough of two-furrow form. It was popular in Scotland, where there was a greater demand for the type of bodies fitted to this plough than the general-purpose and semi-digging types.

This plough continued to be developed into the 1960s, and in 1965 a three-furrow version was introduced, the TS64H, with a further development, the TS64K, with four furrows. By now the Epic body was no longer in general use, and thus only the

Early experiments took place to facilitate ploughing with more than three furrows and this is an experimental four-furrow plough of semi-mounted form. The hydraulics of the tractor were incapable of lifting such a heavy plough, so the rear wheel carried the back of the plough and facilitated raising and lowering to put the rear bodies in and out of work.

SCN and the UDM, plus the bar point bodies, were offered. The last year of its manufacture saw a further change in body type, with the SCN and TCN variants fitted in lieu of the earlier types of digger and semi-digger bodies.

The need to produce a four-furrow plough had led Ransomes to introduce the TS70 and 71 in 1954, both taking either YL or Epic bodies, the latter capable of being fitted with a bar point. The TS70 could cut a furrow up to 10 inches wide, the TS71 up to 12 inches. The frame design was based upon the TS54D and reverted to the arrangement of forged plain swan-

The TS71 was the first fully mounted four-furrow plough developed for the later diesel Majors and seen here with Epic bodies.

The enlargement of the TS59 to encompass four furrows and more led to a reduction in the need to maintain the TS71, which was only a four-furrow type. This is an experimental version of the two which subsequently was put into production as a variant on the TS59 type.

neck beams, unlike the other variants with detachable cast legs. Later that year the TS73 was introduced, a further four-furrow plough with the lighter form of semi-digger body, the FRDCP. The need to provide a lighter body was important, particularly for multi-furrow ploughs provided with a semi-digging facility, which would otherwise be beyond the capabilities of the tractors.

These ploughs were designed for use with the Fordson Diesel Major, and the TS73 was the first of the Ford Ransomes ploughs not provided with a depth wheel, the pre-set linkage facility on the Major not requiring one. The frame design of this plough reverted to that of the TS59, with detachable legs.

The following year, 1955, a two-furrow version of the same plough, the TS72, was introduced and was manufactured for the Ferguson tractor. This plough was developed as a general-purpose plough and was capable of being fitted with YL, Epic, AGT and RND bodies and a bar point provided for the Epic body. It is of interest that this was the first of the Ford Ransomes ploughs not manufactured specifically for a tractor produced by Ford themselves. Their competitor to the Ferguson TE20 and later the 35 range was the Dexta, which was not put into production until two years later, in 1957.

Further development of the reversible plough had continued and a two-furrow export model, the TS68, a replacement for the TS50, was introduced in 1955, which was capable

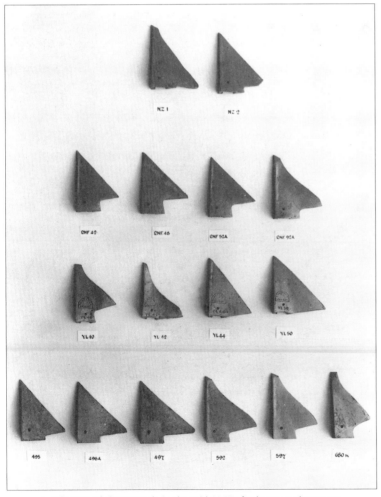

A selection of shares made in the mid 1950s for home and overseas.

of ploughing either 10 inches or 12 inches in width and between 8 inches and 10 inches in depth. This new general-purpose reversible plough was capable of being fitted with the DMC/DMD bodies, both of which could be fitted with bar points and were used for digging work in easy conditions, with YL and Epic bodies being offered for general-purpose and semi-digger work.

The following year a variant, the TS74, was introduced. This was a single-furrow reversible plough primarily for export, fitted with the UDM and NUD deep digging bodies, both of which could be provided with a bar point. It was subsequently developed to take the SCN body, again with bar point as required by soil conditions.

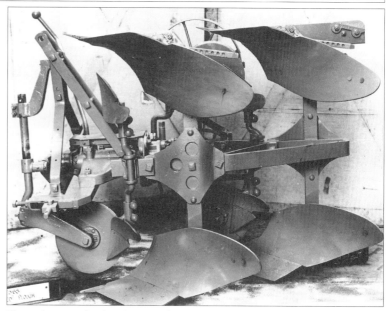

The replacement for the TS50 was the TS68 mounted plough of rather more robust form, seen here with a pair of DM bodies fitted with bar points. This was first used behind the higher horsepowered Diesel Majors. The first models were produced for the export trade.

Ford's long-awaited rival to the Ferguson TE20 and successor, the 35, had arrived in 1957 in the shape of the Fordson Dexta, powered by a three-cylinder diesel engine and similar in size to the Ferguson and its successors. It thus provided Ford with a direct competitor in the smaller tractor market and its introduction led Ransomes to the development of a range of ploughs to fit the new tractor.

The first of these, the TS1013, was introduced in 1957 and was a mounted non-reversible three/two-furrow plough for general-purposes, developed from the older TS54 range. It was capable of being fitted with either the YL, the Epic or the DMD body and was not, of course, provided with a depth wheel, because the tractor was fitted with hydraulics featuring draft control which did not require this feature.

In the same year three other ploughs were introduced. The TS1014 was a one-furrow deep digging plough fitted with either the UDM or the SCN and bar point body for deep and wide work. A plough fitted with the same UDM body, but of reversible design, was introduced at the same time, known as the TS1015. The TS1016 was a two-furrow mounted reversible plough and capable of being fitted with the

A two-furrow TS68 fitted with digger bodies.

semi-digger FRDCP and the general-purpose, YL body. The TS59 was also adapted for use by the Dexta in the early 60s, as discussed on page 71.

The formal arrangements between Ford and Ransome did not of course mean that Ransomes produced ploughs purely for use with Fordson tractors; indeed most contemporary implement manufacturers produced equipment capable of being used with many makes of tractors. Most of the larger tractor manufacturers, however, provided their own range of implements, with the notable exception of Nuffield, who made no implements of their own. They did, however, recommend other manufacturers' items which were a suitable match for their tractors.

Nuffield produced small handbooks showing the approved manufacturers' implements and in the early 1950s the TS50, 51, 55, 59, 63 and 64 ploughs were all recommended along with other manufacturers' products for use with Nuffield tractors. At that time, Ransomes had a Nuffield tractor in the works, and a considerable number of photographs were taken showing implements attached to this tractor and others of this make the company used.

A more formal arrangement was entered into between Nuffield and Ransomes, whereby the latter company produced a specific export catalogue of implements from their general range, including, of course, ploughs available for the overseas markets. These were provided by Ransomes outside the Ford Ransomes range which was being sold in Britain. It is amusing to note that Nuffield's domestic implement brochures do not refer to the ploughs as being of Ford Ransomes manufacture, but merely Ransomes manufacture, for obvious reasons!

The TS1013 was a development of the older TS54 and produced for the Fordson Dexta in 1957. This tractor was fitted with draft control and did not require a depth wheel, and the plough is seen here fitted with the FRDCP semi-digger bodies.

A reversible plough of single-furrow digging form, the TS1015, also available for the Dexta, is seen here fitted with DM bodies. Note the early use of hydraulics for turning the plough.

Ploughs for the Fordson Super Major

At the end of 1960 the Fordson Super Major was introduced, using the same engine as the previous model, the Power Major, which developed just under 52 bhp. This was slightly increased in power later, and advantage could therefore be taken of the improvement in design and updated technology to provide bigger ploughs. In 1961 further development in plough design occurred with the introduction of the TS78, which was of a semi-mounted design, the first such mass-produced plough manufactured by Ransomes.

The TS78 was introduced in the early 1960s and coincided with the development of the "1000" range of Ford tractors. A six-furrow variant is seen here at work. Note the frame body shape and strengthening required, even with the provision of the rear carrying wheel.

This plough was provided with six furrows, but could be converted to five, and was not mounted on the linkage in the conventional manner. The size of the plough would have been far too great for a tractor to lift, and therefore the front of the plough was mounted on the hydraulic lift arms conventionally, but the back was carried by a wheel which was hydraulically actuated by the tractor for the purpose of raising it and lowering it into the ground. The plough was of general-purpose design, and could work up to 8 inches deep, with a range of bodies including YL, FRDCP, Epic with bar point and DMD digger bodies, with a bar point variant if required. It was designed for ploughing large areas not requiring deep

The TS68s were superseded by the TS82 and subsequent ploughs. Here a TS82 fitted with semi-digger bodies is seen at work behind a Super Major.

digging, although not intended as a replacement for the general-purpose TS59, which was then being developed with a four-furrow version to take advantage of the improved power available with the Super Major.

The same year a further development of the mounted reversible two-furrow plough took place in the form of the TS82, which could plough up to 12 inches deep and with a 12 to 16 inches cut. This was available with a variety of bodies, including the YL, Epic, FRDCP, and SCN, and was therefore a general-purpose plough. The following year saw a further development with a three-furrow version introduced, the TS83, which could be fitted with YL, Epic or the DMD digger bodies. Both these ploughs were fitted with a mechanical trip mechanism.

In 1964 the Super Major was superseded by the development of the Ford 1000 Series tractors, the largest of which, the 5000, was capable of producing 67 bhp. During this period the users of Ford skid units, County, Roadless and later Ernest Doe, were developing tractors capable of delivering power on all wheels, which could more usefully plough with the larger ploughs then being developed. These tractors had the facility for accepting mounted implements of more substantial size than had hitherto been available. In 1964 the TS84 and a later variant the TS84A were introduced, comprising a three-furrow digger plough fitted with SCN and bar point bodies capable of ploughing up to 13 inches or 14 inches in width. This was of similar form to the earlier mounted reversible ploughs, but was of course larger in its capacity than anything hitherto produced. All these ploughs were reversed by a manual lever, which was a tiring device to operate when ploughing short fields.

By the middle of the 1960s the range of ploughs

Hugh Barr was an expert champion ploughman who was used by Ransomes for both promotional and experimental testing purposes. He is seen here climbing on to a Super Major prior to beginning work with a TS82.

The development of the larger horsepowered Major tractors in the 1960s led to the development of heavier duty ploughs. Here a three-furrow TS84 is seen at work.

available had been reduced considerably. Of the non-reversible mounted types, the TS59 and TS64 ranges were still available, the TS59 in 2/4- and 5-furrow versions for general-purposes, with a smaller range available for the Dexta tractors with a category 1 linkage. The TS64 plough range continued in 3- and 4-furrow form as the non-reversible mounted plough capable of deep digging, and the TS78 range in five- and six-furrow form as a semi-mounted non-reversible plough for general purposes.

Reversible ploughs existed in the form of the TS82, 83 and 84 as two- and three-furrow general-purpose and digger ploughs, the TS81 being introduced as a two-furrow variant for the Dexta taking category 1 linkage. These, and other ploughs of a similar pattern, represented the third generation of reversible ploughs produced for the home market.

The TS81, introduced in 1960, was a lightweight version of the TS82 designed for use on lighter soils and with smaller tractors.

In the mid 60s the TS86 was made specifically for match ploughing and fitted with modified TCN bodies, hence the provision of the small depth wheel and built-in weight carrier in the centre of the plough. It appears to have been provided with an adjustable top link seen fitted to the head stock.

The fact that these ploughs were developed well over ten years after the introduction of the TS50 and 51 variants shows the very limited extent to which reversible ploughing was then being used in British agriculture. Thus the considerable range of ploughs introduced since 1946 had been reduced to four basic types, and further rationalisation was shortly to occur.

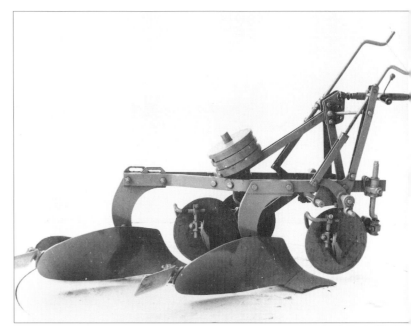

The TS88, made as a special "on land" semi-mounted plough, was used for the important crawler market in the late 60s.

CHAPTER 7

NEW THEME – FINALE

The reduction of plough types in the 1960s was a process that had been occurring continuously over the previous twenty-five years. The diverse range of both mounted and trailing ploughs was now much reduced and body design had been further refined, with a maximum of four types generally being available. The same decade saw the end of trailing plough development and availability, although the TS46 and the larger TS69 Hexatrac continued selling, primarily to those who used crawlers and larger tractors without hydraulics. In 1962, the final development of that type occurred with the introduction of the four-furrow TS85 produced for Holland.

New Theme ploughs

In 1967 a new range of mounted ploughs was introduced which took the place of the TS59 and TS64 variants. This was marketed under the title New Theme and comprised a range of five ploughs, the first of which was the TS90, with two to five furrows for general-purpose ploughing. It was available with the TCN semi-digger body, but could still be fitted with the YL. For the larger ploughs a pneumatic wheel was fitted, providing them with a semi-mounted ability, thus substituting this particular model for the TS78.

It was of radically different design to its predecessors, having a rolled hollow section frame which was a type of steel profile recently introduced by the then British Steel Corporation. The frame was angled to the rear of the tractor and provided with offset legs, which could be fitted with a spring release mechanism in areas where ground conditions could cause potential damage to the plough legs.

In later production, the UCN substituted the TCN; the UCN was itself a variant of the SCN deep digging type, and this body type was used for semi-digging.

The TS90 New Theme was introduced in 1967. It is seen here in five-furrow form fitted with the SCN digger bodies.
The picture was taken in the special photographic studio at Nacton. Note the small size of the turntable which,
within a few years, was too small to take the larger implements.

In 1969, a new range of reversibles was manufactured, replacing those in the TS80 range. Three versions were made initially: the TSR102 was a two-furrow plough for use with medium-powered tractors and provided with general-purpose and semi-digging bodies, including the YL type; the TSR103, a three-furrow variant, took the same bodies, and the TSR106 took an additional furrow and was provided with wheels for running "on land". A further type, the TSR109, also of four-furrow form, took the UCN body and was built for high horsepower tractors. A considerable number of options were available with these ploughs, including land wheels, trip legs, differing skims and discs and, on the larger models, a hydraulic turnover mechanism.

The range was extended in 1973 with the introduction of the TSR107 and 108, which were two- and three-furrow developments. Progressively over the next few years further versions appeared, taking similar bodies but with specific ploughs being

It could be provided with a spring-loaded bar point if required and this offered additional protection against damage from underground obstructions.

The TS94 and TS95 were introduced later, comprising four/six- four/five-furrow versions respectively, capable of ploughing at different widths and depths to the TS90. The TS95, first made in 1969, was capable of being used for both deep digging and general purposes as required, offering a greater range of applications than the preceding models. Later the TS96 was introduced, which was an "on land" version of the same plough, capable of being used by higher horsepower four-wheel drive tractors and crawlers.

The TSR108 was a large capacity three-furrow plough with large under-beam and inter-body clearances. designed to deal with trashy conditions; the plough required the rear wheel for additional support. All the ploughs in this range were a development of the TS82 and similar reversibles adapted to the newer bodies then being developed.

The final development of the non-reversible mounted plough was the Spaceframe TS200 range. Capable of taking all the combinations of bodies Ransomes produced, the plough was also available as a long or short version, offering a variety of inter-body clearance. It was even available as a small three/two-furrow version for mini-tractors that were popular for small farms and smallholdings at the time.

introduced with fixed widths and body types. The provision of disc coulters was not universal at this time and trash accumulation had been a chronic problem, which was overcome by the omission on some models of any form of coulter. A special model, the TSR112, had been introduced for rock-strewn soils, and provided with a hydraulic turnover and fitted with the bar point bodies to deal with adverse conditions. All these ploughs were similarly constructed, with rolled hollow steel section frames set at an angle to the rear of the tractor and provided with headstock arrangements of a similar sort to those introduced previously.

This decade saw a substantial amount of development in the design of reversible ploughs, and Ransomes must have been busy undertaking the necessary research and design to introduce such a wide range of differing ploughs.

In 1976, the final development of the non-reversible mounted plough came with the Spaceframe 200 series. This comprised a range of ploughs which was reputedly of radically different design, as Ransomes claimed that they had produced a new plough from the ground up in terms of design features. In practical terms the construction was not radically different in concept, but in terms of ability to deal with a wide range of differing soil conditions it produced the maximum amount of flexibility from the same basic plough design.

The series was available initially in two different forms: the short clearance version had 30½ inches between bodies and was for between two and six furrows; the long clearance had 40½ inches and was for two to five furrows. Like its predecessor it could be provided with a rear wheel, making it semi-mounted in form. It was also available with legs with shear bolts to prevent damage, or a spring trip mechanism of a sort similar to that of its predecessor, the TS90 series, or alternatively, a more expensive and complex hydraulic trip leg. The latter would only have been used in areas where breakage was a considerable problem and where there was a need to maintain high speed ploughing under adverse ground conditions. The plough was also provided with hydraulic furrow width adjustment, but retained mechanical cross shaft adjustment, although with the option of hydraulic adjustment if required. It could also be provided with a depth wheel, if this were necessary for precision ploughing and with it a choice of three skims and discs. The bodies were the YL, UCN and SCN variants, and a bar point body was available for stony ground conditions.

The 1960s had seen the dominance of the reversible plough grow to the extent that one suspects that sales, both actual and potential, for a mounted plough of the non-reversible sort were much reduced. All the major manufacturers were producing a similar sort of plough at this time, when scientific research on alternative weed control measures had begun to question the very need for ploughing at all. It was also

not directly sold by Ransomes, but were marketed by them in view of the fact that they had produced nothing of a similar sort for this small and specialised market. The plough types involved were: the RDP1, which was a three- to six-furrow variant taking general-purpose, multi-purpose and deep digging bodies, including the older UDM type body; the RDP2, a five- to nine-furrow variant taking the UDM deep digging body; and the RDP6, a four- to six-furrow version taking the same body.

Three years later, in 1980, Ransomes introduced the final version in the TSR100 reversible series, the TSR113 and 114, which were general-purpose ploughs produced for what was called non-stop ploughing. There were three- and four-furrow variants taking the general-purpose body and provided with sprung reset legs. They were specifically designed for high output ploughing, where there was a need to continue ploughing at some speed, and with the availability of a spring re-set on the legs, ploughing could continue in the knowledge that it was unlikely that damage would delay progress.

That same year a further plough, the TSR150, in five- to seven-furrow form, was introduced, specifically for high horsepower tractors and incorporating a complete range of body types, from general-purpose to deep digging and including bar point versions as has been described previously. Subsequently, the TSR250 was introduced, a similar plough, but this time capable of being provided with further furrows, and produced with the same types of body, with the addition of the RCN variant, available for the export market only.

Development continued apace however, and in

a decade in which there was considerable economic upheaval, which had a significantly adverse effect upon the fortunes of all major manufacturing companies. It must have put Ransomes under some strain in terms of development costs, let alone manufacturing, and these reversible ploughs, in terms of sophistication and complexity, were beyond anything that Ransomes had produced previously. The much greater costs of introducing such ploughs, compared to the simple ploughs of only ten or fifteen years previously, greatly increased their selling price. Furthermore, taking into account the much higher performance and utilisation of these larger multi-furrowed ploughs, there would have been significantly less potential demand.

In 1977 the firm Dowdeswell, who were at that time a small specialised plough manufacturer, had produced a range of ploughs with category 3 linkage for the large crawler market. Dowdeswell utilised Ransomes' bodies in a way similar to other commercial arrangements in which Ransomes had been engaged for many years. These ploughs were

The TSR300 range represented the final development of the mounted plough. This is the basic TSR300D four-furrow plough.

1982 Ransomes introduced their final range of ploughs, the TSR300 series, provided with the same range of bodies as for previous ranges, but excluding the YL pattern, and capable of being provided with two to six furrows. They were of similar construction to the preceding ranges, and had a wide variety of options of similar sort, with skims, depth wheels and such features available. This entire range was provided with hydraulic reversing mechanism for the first time, whereas preceding ranges only had this feature on the larger ploughs.

A variety of different features were available in this range, giving the widest possible choice depending upon soil conditions. The standard two/four-furrow plough was designated the 300S, with the four- or five-furrow version the HS, where higher horsepower tractors were required. The 300D was capable of being fitted with disc coulters for grassland ploughing, and the frame was longer to accommodate this feature. The 300LD had an even longer frame and cutaway bodies to deal with trash and straw. The 300HD was a five- or six-furrow version, to deal with ploughing big acreages, and the 300BLS a five-furrow version fitted with extra long legs to plough at the maximum width of 14 inches or 16 inches. The 300T was fitted with trip legs, to allow non-stop ploughing without interference by having to stop to clear blockages or obstructions. The 300LT was a version fitted with a steel depth wheel and scraper with optional pneumatic tyre depth/transport wheel.

A particularly interesting development was the TSR300FD, which was mounted on the front of the tractor and actuated by a special linkage; Ransomes providing the mounting plate for attachment. This was available with a similar range of bodies as those provided on other ploughs, and was, of course, intended to be utilised with a rear multi-furrow plough. In practical terms it does not appear to have been a great success. The additional number of furrows provided was offset by the difficulty in setting the plough up correctly before work started and by the attendant problem of ensuring that both front and rear ploughs were working correctly, thus putting additional strain upon the tractor driver's concentration.

This range of ploughs continued in production for a short period to use up existing stocks of components left after the takeover of Ransomes' farm machinery business by the Agrolux Company. Ransomes had been designing another range of reversibles, the TSR 500, but these never left the drawing board.

The TSR300S four-furrow reversible had closely set bodies and was available with a choice of under-beam clearance.

TSR300HD with TSR300FD front mountable reversible shown in action. The front-mounted plough proved to be a relatively short-lived phenomenon.

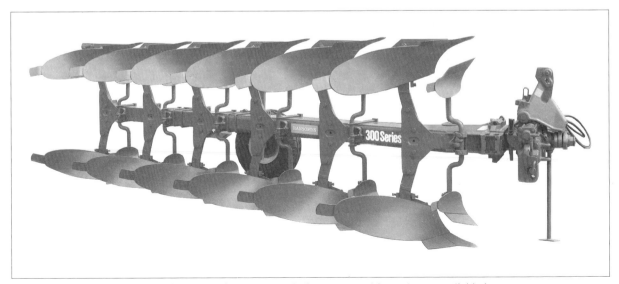

A TSR300HD in six-furrow configuration made for tractors with maximum available horsepower. A five-furrow long clearance version was also available.

POSTSCRIPT

The cessation of the plough manufacture by Ransomes has not meant the end of ploughing with Ransomes' ploughs. The ruggedness of Ransomes' engineering will ensure that it will be, no doubt, many years before the last of the products of Orwell and Nacton turn a furrow for the last time. There is, in any case, another role in which their products will be appearing for many years, and that is in match ploughing. The yearly growth in popularity of that activity, particularly with the use of vintage tractors, sees probably as many ploughs put into action for that purpose as are taken out of commercial use. The types used are predominantly those that take the general-purpose bodies, such as the RSLDs and Motracs, for trailing purposes, and the TS59s, TS63s and the TS54 Robin. With the advent of post-1960 Classic tractors being preserved in larger numbers, late model TS59s and TS90s are moving from contemporary classes to those representing an earlier era. The large numbers of Fordson ploughs made guarantee that representatives of the earlier period of collaboration with Ransomes are also seen competing in the many matches that take place during the ploughing season.

Unfortunately, digger ploughs, especially the deep digging varieties, rarely appear in such competitions, and are only likely to be preserved in museums or by those who have ground of their own to plough. The sheer size of the later reversible ploughs makes their preservation more of a problem, especially in transporting them to ploughing matches. In addition there are the problems of preserving the modern tractors, with which to use them. Apart from their size, the electronic sophistication of the tractors will make them a daunting prospect for the preservationists and restorers of the next generation.

Many connoisseurs of the trailing classes still consider that the work done by an RSLD, with Newcastle bodies and hauled by a Standard Fordson in the hands of a skilled ploughman, represents the ultimate in standards of workmanship. The number of ploughs of this era that remain to be discovered are probably very limited, and the chances of finding them fitted with long boards must be slim indeed. A good alternative would be a YL body, a YL165, although good used boards for these ploughs are an increasing rarity, but new 183s are being made again. A further problem that faces the vintage ploughman is the difficulty in finding complete ploughs. Many of those previously relegated to secondary duty on the farm or to the scrap heap will lack skim coulters and possibly disc coulters. These parts are now harder to find, and at some time in the not too distant future someone will perhaps consider manufacturing such components again, although shares, discs and recently mouldboards are still being made by a few specialist engineers. If only Ransomes could have seen this renaissance before the end of manufacture, perhaps steps could have been taken to ensure continuity of some component manufacture in the longer term.

This book is written more than ten years after Ransomes sold their agricultural business and stopped making ploughs, and the world of agriculture faces changes of a more radical form than for perhaps a generation and a half. Machinery sales of all sorts are in decline, with further contraction in well-known manufacturing names, and probably more to come. Ransomes was perhaps the greatest name in British-owned agricultural engineering companies and the only one that was pre-eminent in various fields of manufacture worldwide. Perhaps after all, their move from agriculture to concentrate on turf care machinery manufacture will prove to be a wiser move than others thought at the time it occurred.

GLOSSARY

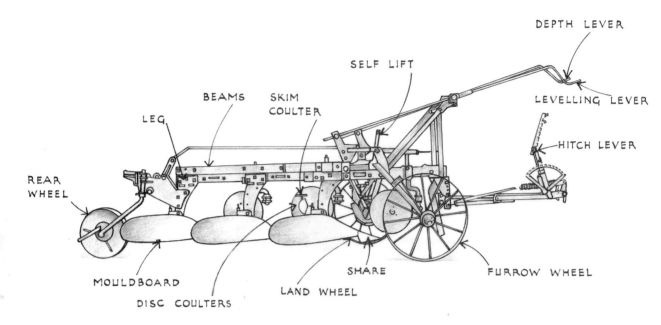

DEPTH LEVER

SELF LIFT

BEAMS   SKIM COULTER

LEG

LEVELLING LEVER

HITCH LEVER

REAR WHEEL

MOULDBOARD

DISC COULTERS

LAND WHEEL

SHARE

FURROW WHEEL

Breast	*See mouldboard.*	**Frog**	The malleable or steel casting or fabrication which is fixed to the bottom of the plough leg and provides the support for the mouldboard. Sometimes referred to as a saddle.
Cross shaft	The shaft which, by revolving using a control handle, turns the front body relative to the front and rear axis of the plough, and either narrows or widens the furrows turned.		
		Furrow wheel	The wheel that partially carries a trailing plough and runs in the furrow.
Depth wheel	The wheel whose adjustment vertically by one of the control handles alters the depth of soil turned by the plough.	**Knife coulter**	A fixed blade shaped component, which undertakes the same function as the disc coulter *(see above)*.
Disc coulter	A rotating disc which is fixed to the frame of the plough on a vertical arm and which, running ahead of the share, cuts a vertical slot in the soil against which the share starts the horizontal cut to form the bottom of the furrow.	**Land wheel**	The wheel that partially carries a trailing plough and runs on the land.
		Landside	A flat steel or iron plate fixed to the frog that runs against the furrow wall and provides a flush finish to same.

Mouldboard   The curved plate fixed to the frog behind the share that turns the soil to create the furrow. The shape of the mouldboard gives the characteristic form to the furrow, and can be rounded, partially rounded or broken in appearance.

Mounted plough   A plough carried entirely on the hydraulic lift arms on the back of the tractor.

Non-reversible plough   A plough that turns the furrow in one direction only, and requires the field to be set out in lands or areas to make economic use of the tractor.

Plough body   The assembled share, mouldboard, frog and supporting brackets.

Reversible plough   A plough with two sets of bodies that turn alternative left and right furrows, and allows ploughing to be undertaken continuously without the tractor having to plough in a pattern within the field to make economic use of its power. It saves in not having to spend time marking out the field before starting to plough and only requires the setting out of the head and side land.

Semi-mounted plough   A plough of a size too large to be entirely mounted on the rear of the tractor and therefore partially carried on a rear wheel.

Skife   The detachable leg of a plough that is bolted to the frame of the plough with a moulded shape for direct connection of the mouldboard. Does not require the use of a separate frog for the same purpose.

Skim coulter   A small shaped casting that is fixed above the mouldboard to cut a corner off the furrow as it is being inverted, and allows the slice to lie without any vegetable matter being left exposed on a corner of the upturned soil.

Swan-neck beam   A beam with one end forged with a curved shape to form the leg of the plough.

CLASSIFICATION OF
RANSOMES TRACTOR SHARE PLOUGHS

Code or No.	Name	Introduced	Type	Furrows	Body types	Purpose
TS1	Solotrac	1931	Tr.N.R.	1	UN, UG, UD + BP	Digging
TS2	Proconsul	1932	Tr.N.R.	2,3,4,5	DM + BP	Export
TS3	Proconsul	1932	Tr.N.R.	2,3,4,5	DM + BP	Export
TS4	Giantrac/Jumbotrac	1931	Tr.N.R.	2,3	UN, UG, UD	Digging
TS5	Motrac	1931	Tr.N.R.	2,3	DM + BP	Export - Italy
TS6	Duotrac	1931	Tr.N.R.	2	UD + BP	Digging
TS7	Quintrac	1931	Tr.N.R.	5	DM	Export
TS8	Unitrac	1932	Tr.N.R.	1	JUM	Digging
TS9	Motrac	1936	Tr.N.R.	4	DM	Export - Holland & Italy
TS10		1932	Tr.N.R.	4	UN, UG, UD	Export
TS11	Junotrac	1933	Tr.N.R.	2,3	UD	Export
TS12	RSLD/M	1933	Tr.N.R.	2,3	YL	Orchard work
TS13	Junotrac	1933	Tr.N.R.	3,4	UD	Export
TS14	Midtrac #		Tr.N.R.	3	DM	Export orchard
TS15	Unitrac	1934	Tr.N.R.	1	UN	Home & export
TS16	Unitrac Major/Estotrac	1934	Tr.N.R.	1	JUM, SL	Digging & export
TS17	Junotrac	1934	Tr.N.R.	3	YL	Export
TS18	Marquis	1934	Tr.N.R.	3,4	UN	Export
TS19	Marquis	1934	Tr.N.R.	4,5	UN	Export
TS20	Hexatrac	1937	Tr.N.R.	5,6	LCP	Digging
TS21	Junotrac	1934	Tr.N.R.	2	YL	Export
TS22	Hexatrac	1934	Tr.N.R.	5,6	YL	General purpose
TS23	Magnatrac	1936	Tr.N.R.	3,4	JUM, UN	Digging
TS24	Twinwaytrac	1935	Tr.N.R.	2	UD, SCPT	Digging
TS25		1935	Tr.N.R.	2	Horticultural	General purpose
TS26	Motrac Major	1936	Tr.N.R.	4	YL	General purpose
TS27	Unitrac Minor	1936	Tr.N.R.	1	UN	Digging
TS28	Junotrac	1936	Tr.N.R.	4	UN	Export
TS29	RSLD/M Major	1936	Tr.N.R.	2,3	YL, IRDCP	General purpose
TS30		1937	Tr.N.R.	I	RHA, EC	MG Crawler
TS31		1937	Tr.N.R.	2	VY, SHP	MG Crawler
TS32	Proconsul	1938	Tr.N.R.	4		Export - Holland
TS33	Twinwaytrac	1935	Tr.N.R.	1	SCPT	Digging
TS34	Marquis	1935	Tr.N.R.	2,3	UN	Export
TS35	RSLD	1939	Tr.N.R.	2	YL	Special for hop fields
TS36	Twinwaytrac Minor #	1939	Tr.N.R.	1	TCP	Digging
TS37	Proconsul	1940	Tr.N.R.	3	DMC	Export
TS38	R.D.S.	1940	Tr.N.R.	2,3,4	TCP, SCP, YL, IRDCPT, R&H	General purpose home & export - Australia
TS39	No. 7 Hexatrac	1940	Tr.N.R.	5,6	LCP, YL	General purpose
TS40	No. 10 Quintrac	1940	Tr.N.R.	5	LCP, YL	General purpose
TS41	Supertrac	1947	Tr.N.R.	4	UN, UG, SU	Export
TS42		1945	Tr.N.R.	1,2	RHA, VY, EC SHP, YL, IRDCPT,	MG Crawler
TS43	Motrac	1946	Tr.N.R.	2,3	YL, IRDCPT, TCP, LCP, GT	General purpose
TS44	Litrac	1946	Tr.N.R.	2,3	DMC	Export
TS45	Midtrac	1948	Tr.N.R.	2,3	YL, Epic, GT, LCP, TCP	General purpose
TS46	Multitrac	1948	Tr.N.R.	3,4	YL, Epic, GT, LCP, TCP, UD	General purpose
TS47	Duratrac	1947	Tr.N.R.	3	UD, UD-BP	Export
TS48	Duratrac	1948	Tr.N.R.	4,3	UD, UD-BP	Export
TS49	Duratrac	1949	Tr.N.R.	5,4	UD, UD-BP	Export
TS50		1949	M. Rev.	2	YL, Epic	General purpose
TS51		1949	M. Rev.	1	UD, UD-BP	Digging
TS52	RSLD/M#	1949	Tr.N.R.	2,	YL	Special for hop fields

Code or No.	Name	Introduced	Type	Furrows	Body types	Purpose
TS53		1950	M. N.R.	1	UD, UD-BP.	Export
TS54	Robin*	1950	M. N.R.	1,2,3	Epic, YL, DMD	General purpose
TS55	Falcon*	1951	M. N.R.	1	UDM , UDM-BP	Digging
TS56	Protrac	1950	Tr.N.R.	4,3	DM	Export
TS57	Pretrac	1950	Tr.N.R.	5,4	DM	Export
TS58	Midtrac Major	1950	Tr.N.R.	4,3	Epic, YL, GT	General purpose
TS59	Raven*	1951	M. N.R.	3,4,5	Epic, YL, IRDCP, EFR, FRDCP, TCN	General purpose
TS60	Not used					
TS61		1950	Tr. N.R.	1,2,3	VY	Market garden - Australia
TS62		1952	M. N.R.	2	UDM, UDM-BP, DMD-BP	Export
TS63		1951	M. N.R.	2	Epic, YL, GT, RND, RND-BP	General purpose
TS64		1951	M. N.R.	2,3,4	Epic, SCN, UDM, UDM-BP	Digging
TS65		1951	M. N.R.	1 Right hand	SHPM	MG Crawler
TS66		1951	M. N.R.	1 Left hand	SHPM	MG Crawler
TS67		1951	M. N.R.	2	VYM, RJS	Export MG Crawler
TS68	Swift*	1955	M. R.	2	DMC, DMD-BP, YL, Epic	Export & home
TS69	Hexatrac	1952	Tr. N.R.	6	Epic, YL	General purpose
TS70		1954	M. N.R.	4	Epic, IRDCP, YL	General purpose
TS71		1954	M. N.R.	4	Epic, IRDCP, YL	General purpose
TS72		1954	M. N.R.	2	Epic, YL, GT, RND, RND-BP	General purpose
TS73	Heron*	1954	M. N.R.	4,3	FRDCP, YL	General purpose
TS74	Swallow*	1954	M. R.	1	UDM, UDM-BP	Digging
TS75		1954	M. N.R.	1	SCN-BP, UDM, NUD-BP	Digging
TS76	Not used					
TS77	Not used					
TS78		1961	S.M. N.R.		YL, FRDCP, Epic-BP, DMD-BP	General purpose
TS79	Not used					
TS80		1960	M. R.	1	UDM, SCN SCN-BP	Digging
TS81		1960	M. R.	2	YL, TCN, UCN , TCN-BP	General purpose
TS82		1960	M. R.	2	YL, TCN, UCN, TCN-BP, SCN, SCN-BP	General purpose
TS83		1960	M. R.	3	YL, TCN, UCN, TCN-BP	General purpose
TS84		1964	M. R.	3	UCN, SCN, SCN-BP	Digging
TS85		1964	Tr. N.R.	4	UDM, SCN, SCN-BP	Export
TS86		1965	M. N.R.	2	Epic, TCN + Special body	Match ploughing
TS87		1965	M. N.R.	2,3	SCN, Epic	Match ploughing
TS88		1967	S.M. N.R.	5	SCN, SCN-BP	
					TCN, TCN-BP	Digging (Crawler)
TS89		1968	M. R.	4	YL, TCN, UCN, TCN-BP	General purpose
TS90	New Theme	1967	M. N.R.	2,3,4,5	YL, UCN	General purpose
TS91	New Theme	1967	M. N.R.	2,3,4	UCN, UCN-Y-BP,	
					SCN, SCN-Y-BP	Digging
TS92	New Theme	1968	M. N.R.	6,7,8	YL, UCN	General purpose
TS93	New Theme	1968	M. N.R.	4,5,6	YL, UCN	
TS94	New Theme	1968	M. N.R.	4,5,6	UCN	
TS95	New Theme	1968	M. N.R.	4,5	UCN, UCN- -BP, SCN, SCN-Y-BP	Digging
TS96	New Theme	1969	M. N.R.	7,6 & 6,5	YL, UCN, SCN, SCN-Y-BP	General purpose for "On" and "Off" land
TS97	Ransomes Bonning	1979	M. N.R.	2	Bonning	Match plough
TS200	Spaceframe	1976	S.M./ MN.R.	2,3,4,5,6	YL, UCN, SCN, SCN-BP	General purpose
TS1013	Robin*	1957	M. N.R.	3,2	YL, FRDCP, DMD	General purpose

Code or No.	Name	Introduced	Type	Furrows	Body types	Purpose
TS1014	Tern*	1957	M. N.R.	1	UDM, SCN-BP	Digging
TS1015		1957	M. R.	1	UDM, SCN-BP	Digging
TS1016		1957	M. R.	2	YL, FRDCP, DMD	General purpose
TSR102		1972	M. R.	2	YL, SCN UCN-Y-BP, SCN-Y-BP, TCN-BP	General purpose
TSR103		1972	M. R.	3	YL, UCN, TCN-BP	General purpose
TSR106		1972	M. R.	4	UCN	Digging "On" land
TSR107		1973	M. R.	2	UCN, SCN, SCN-BP	General purpose
TST108		1973	M. R.	3	UCN, SCN-Y-BP	General purpose
TSR109		1975	M. R.	4	UCN, UCN-Y-BP	General purpose
TSR110		1975	M. R.	3,4	UCN, UCN-Y-BP	General purpose
TSR110/12/3		1975	M. R.	3	YL, UCN	General purpose
TSR110/12/4		1975	M. R.	4	YL, UCN	General purpose
TSR111		1979	M. R.	3,4	YL, UCN	General purpose
TSR112		1977	M. R.	3,4	UCN, SCN, UCN-F-BP, SCN-F-BP, SCN-Y-BP	General purpose. Special for rock-strewn soils
TSR113/114	Non-stop	1980	M. R.	3,4	SCN-D	General purpose. Reset legs
TSR150		1980	S.M. R.	5,6,7	YL, UCN, SCN, UCN-F-BP, SCN-F-BP, UCN-Y-BP, SCN-Y-BP	General purpose
TSR250		1980	S.M. R.	5,6,7,8	YL, UCN, SCN, UCN-F-BP, SCN-F-BP, UCN-Y-BP, SCN-Y-BP	General purpose
TSR300S		1982	M. R.	2,3,4	UCN-D, SCN-D, UCN-Y-BP, SCN-Y-BP, YCN, SLT	General purpose For close coupled work
TSR300HS		1982	M. R.	4,5	UCN-D, SCN-D, YCN, SLT.	General purpose. Stronger version of S model
TSR300D		1982	M. R.	2,3,4	UCN-D, SCN-D, UCN-Y-BP, SCN-Y-BP, YCN, SLT	General purpose. Extended front frame
TSR300LD		1982	M. R.	2,3,4	UCN-D, SCN-D, UCN-Y-BP, SCN-Y-BP, RCN, SLT	General purpose. Long clearance model
TSRHD300		1982	M. R.	5,6	UCN-D, SCN-D, YCN, SLT.	General purpose. A 5- and 6- furrow version of the basic plough
TSR 300BLS		1982	M. R.	5	UCN-D, SCN-D, UCN-Y-BP, SCN-Y-BP, SLT	General purpose. A long clearance version of the HD
TSR300FD		1982	M. R.	2,3	UCN-D, SCN-D, UCN-Y-BP, SCN-Y-BP, SLT	General purpose. A front mounted 3- furrow plough
TSR300FT		1982	M. R.	2,3	UCN-D, SCN-D, UCN-Y-BP, SCN-Y-BP	General purpose. Front mounted plough with spring operated trip legs
TSR300T		1982	M. R.	2,3,4	UCN-D, SCN-D, UCN-Y-BP, SCN-Y-BP	General purpose. Fitted with sprung loaded re-settable trip legs.
TSR300LT		1982	M. R.	2,3,4	UCN-D, SCN-D, UCN-Y-BP, SCN-Y-BP, RCN	General purpose. As 300T but with long clearance between legs

Tr. = Trailing   M. = Mounted   R. = Reversible.   N.R. = Non-reversible   S.M. = Semi-mounted.   * = Export name.   # = Not put in production.
NOTE Subsidiary letters and numbers omitted from descriptive marks.

INDEX

Numbers in bold type refer to illustrations

Agrolux Company 11, 85
Akester, W.D. (Bill) **62**
Ayers, Fred **62**
Barr, Hugh **78**
Black, John 60
Booth MacDonald 29
Brittain garden tractor 47
Byewater, Jim **61**
Churchill tank 19, **20**
Cocksedge & Co. Ltd 20
Cox, Jack 57, 59
Deck, Henry 20, **61, 62**
Deere, John 16
Dowdeswell Engineering Ltd 17, 84
Dyer, Fred 7, 20
Eckert 33
Eden, Sir Anthony **63**
Electrolux 11
Ferguson, Harry 42, 55, 57, 60
Ferguson Brown tractor 42
Ferguson tractors 55, 60, 74
Ford, Henry 38, 57
Ford Motor Company 6, 10, 15, 18, 55-80
    Leamington Spa (Imperial Foundry) 55, 57, 59
Ford Ransomes ploughs
    Ford Ransomes Mounted Plough 6
    FRDCP - Double Chilled Plough 25, 30
    FR EP Plough 64, 67-8, **68**, 69, **colour section**
    FR PM Plough **60**, 61, **62, 63**, 64, **64**, 67, 69, 70
    See also entries for TS50/51, 53, 54, 55, 59, 63, 64, 68, 70, 71, 72, 73, 74, 78, 81, 82, 83, 84, 86, 88 and 1013-16 under Ransomes Ploughs
Ford Tractors
    Ford 8N 65
    Ford 1000 series 78
    Fordson Dexta 10, 11, 20, 74, 75, 79
    Fordson Diesel Major 15, 74
    Fordson Major 6, 18, **45**, 55, 56, 59, **65**, 67, **68, 70**
    Fordson Model F 38, 42, **43**
    Fordson Standard 56, 87
    Fordson Super Major 76, 77, 78, **78**
Fordson Elite Plough 18, 55-6, **56, 58**, 59, 60
Foxwell, John 55
Gass, Jimmy 33
Hunt, Reuben **61**
International Harvester 10/20 tractor 42
Ivel Tractors 34
Jarvis, H. **61**
Johnson, George 14, 30

Kristeel 16-17, 29
Lane, John 16
Lees, Bill 57-8
Martinelli 32
Nuffield Tractors 76
Power, Harry **61**
Ransome, David 32
Ransome, Robert 9
Ransomes, Sims and Jefferies 9-23
    Nacton works 10, **11**, 20, **21**, 23, **58**, 59, 69
    Orwell works 9, **10, 12**, 20, 57, 66
    St. Margaret's Ditches 9
Ransomes Budding machine 9
Ransomes C57 mounted cultivator **60**
Ransomes publications
    Good Ploughing 22
    Brochures 22-3
Ransomes Share Ploughs
    Body types
        AGT 30, 72, 74
        BP - Bar Point 32, 84
        Bonning 27
        DM - Demon 31, **31**, 34, 48, **49**, 50, 53, 69, 75, **75, 76**, 77, 78
        DTP - Deep Tractor Plough 34, **35**
        EC - Eckert Cape 33, 47
        EFR - English Ford Ransomes 31, 61
        Epic **15**, 30, **31**, 34, 40, 61, 64, 65, 66, 69, 70, 71, 72, **72**, 73, **73**, 74, 75, 77, 78
        Farmer's Deck 9, **20**
        FRDCP 71, 74, 76, 76, 77, 78
        GT **31**, 40, 44, **45**
        Guidtop 30
        IRDCP - Irish Ransomes Double Chilled Plough 25, 30, **31**, 33, 34, 40, **44**, 44, 45, 47, 65
        JUM 32, 50
        KCN 33
        LCP - Lincs Chilled Plough 25, 30, **31**, 34, 40, 42, 44, 45, 50
        MCN 33
        NUD 32, 75
        RCN 33
        RHA 34, 47
        RMTM 36
        RMTD 36
        RN - Ransomes Newcastle 9, 24, 30, 34, 87
        RND 40, **40**, 72, **72**, 74
        RSA - Ransomes South Africa 34, 47
        RTPYL 37
        RTPTCP 37

RYLT	36-7, **36,** colour section
SCN	32-3, 34, 72, **72,** 73, 75, 78, 83, 84
SCP - Steel Chilled Plough	25, 30, 34
SCPT	32, 47
SHP - Smallholder Plough	33, 47
SL - Swampland	32, 50
SLT	33, 34
SU	32, 52, **52**
TCN	**32,** 33, 34, 70, 73, **80,** 81
TCP - Titt Chilled Plough	24, 25, 30, **31,** 34, **39,** 40, 42
UCN	33, **33,** 34, 81, 82, 83, 84
UD	32, **46,** 47, 51, 53, 66
UDM	71, 72, 73, 75
UG	51, **51,** 53
UN	32, 51, 53
VY - Victory	33, 34
YCN	33, 34, 47, 84
YOL - Yorkshire Heavy	24
YL - Yorkshire Light	9, 19, 24, 28-30, **29,** **31,** 33, 34, 37, 39, 40, **41,** 42, **43,** 44, **44,** 45, 47, 51, 61, 65, 66, 69, 70, 71, 73, 74, 75, 76, 78, 81, 82, 83, 87, colour section
YLTM - YL Tractor Multiple	34

Code or No.
TS1	26, 51
TS3	49
TS4	26, 51-2
TS5	49
TS7	45
TS10	**50**
TS12	41
TS14	44
TS19	51
TS24	20, **46,** 47
TS25	34, 47
TS31	47
TS33	47
TS35	41
TS38	42
TS30	34, 47
TS41	52-3, **52, 53**
TS42	47, **47,** colour section
TS43	26, 44, **44,** 50
TS44	26, 50
TS45	14, 44
TS46	32, 45, 53, 81, colour section
TS47	27, 53
TS48	27, 53
TS49	27, 53
TS50/51	15, 21, 65, 66, 67, **67,** 76

TS53	**71**
TS54	**21,** 27, 31, 65, 67, 87
TS55	27, 71, 76
TS58	45, **45**
TS59	**15, 21,** 26, 27, 67, 68-71, **69, 70, 71, 74,** 76, 79, 87
TS63	34, 67, 71, **72,** 76, 87
TS64	27, 32, 72-3, **72,** 76, 79
TS65/66	48, **48**
TS68	21, 27, 74-5, **75**
TS69	46, 81
TS70	73
TS71	73-4, **73**
TS72	34, 74
TS73	27, 74
TS74	27, 75
TS78	16, 76, **77, 79**
TS81	79, **79**
TS82	33, **77,** 78, **78,** 79
TS83	78, 79
TS84	78, **79,** 79
TS85	53, **54,** 81
TS86	22, **80**
TS88	**80**
TS90/91	27, 81, **81, 84,** 87
TS94-96	82
TS97	22, 27
TS200	27, 83, **83**
TS1013-1016	26, 27, 75, **76**
TSF200	**84**
TSR100	27
TSR102-109	82, **82**
TSR112	83
TSR113-4	84
TSR150	84
TSR250	84
TSR300	27, 33, 34, 85, **85, 86,** colour section

Name
Ambassador	27, 49
Autocrat	27, 48
Consul	27, 48
Dictator	27, 48
Deep Furrotrac	50
Deeptrac	50
Duotrac	25, 32, 51
Duratrac	26, 53
Emperor	37
Falcon	27
Giantrac	51-2
Heron	27
Hexatrac	25, 45, **54,** 81
Jumbotrac	21, 26, 32
Junotrac	27, 52
Litrac	26, 50
Magnatrac	53
Marquis	27, 51
Midtrac	14, 25, 44, **44, 45**
Midtrac Major	45, **45**
Monotrac	**52**

Motrac	13-14, 18, **18**, 25, 26, 43-46, **43**, **44**, **49**, 57, **60**, 61, **61**, 87
Multitrac	14, 21, 25, 26, 32, 45, 53, **54**
NB	47
New Theme Ploughs	81-6
PE165	**46**
PE516	**13**
PE846 PE	64, **65**
Proconsul	27, 49, **49**, 53
Quintrac	25, 45
RAPT	51
RCLD/M	39
RDS-Ransome Digger Scotland	42
RSLD/M - Ransomes Self Lift Double/Multiple	12-13, **16**, 18, **22**, 23, 24, 38-41, **38**, **39**, **40**, **41**, 50, 57, 67, 87, **colour section**
RSML	24
RST - Ransomes Sub-soiler Tractor Plough	41
Raven	27
Rey	27
Robin	27, 31, 65, 87
SC/Riffler	46
Solotrac	26, 27, 51, **51**
Spaceframe	27, 83, **83**, **84**
Supertrac	27, 32, 52, **52**, 53

Swallow	27
Swift	27
Tern	27
Twinwaytrac	20, 32, 46, **46**
Unitrac	25, 27, 31, 32, 34, **50**, **50**, 51
Vice Consul	27, 49
Vulture	27
Weetrac	25, 42, **42**, **43**
Ransomes' Tractors	
MG range	34, **35**
MG2	33, 47, **47**
MG5	**48**
MG6	33, 47
Ronayne, Mick	**61**
Rotherham Forge	16
Royal Agricultural Society	28, 39
Sankey	15
Saunderson	34
Sheffield Forge	16
Skinner, Ted	59, 60
Standard Motor Company	55, 60
Steel Case Manufacturing Company	16-17, 33
Stormont Engineering	29, 69
Teague, Geoff	21
Textron Golf, Turf and Speciality Products	11
University of Reading, Rural History Centre	6, 26

ANTHONY CLARE

Tony Clare is a Chartered Building Surveyor from Surrey whose lengthy vacations on a family farm near Salisbury in Wiltshire inspired a lifelong interest in agricultural history and in particular machinery. He has a small collection of vintage agricultural equipment including two tractors, both of which see action in the autumn at vintage ploughing matches using Ford Ransome ploughs.

This is the first book he has written on an agricultural history topic, although he has undertaken private research on various topics over the years. The book, which provides an historical analysis of Ransomes and the tractor share ploughs they manufactured, has grown from what was intended to be short-term research at the Rural History centre at Reading University, where the Ransomes archive is held.

Ran